Metamorphosis

Metamorphosis

How insects are changing our world

Erica McAlister with Adrian Washbourne

PUBLISHING

I dedicate this book to my mum who listened and laughed
at my scribbles. Thank you for being the best that there was.
Erica x

First published by the Natural History Museum, Cromwell Road, London
SW7 5BD © The Trustees of the Natural History Museum, London, 2024

A catalogue record for this book is available from the
National Library of Australia

ISBN 9781486318902

This edition published exclusively in print only,
in Australia and New Zealand, by:
CSIRO Publishing, 36 Gardiner Road, Clayton VIC 3168, Australia
Telephone: +61 3 9545 8400
Email: publishing.sales@csiro.au
Website: www.publish.csiro.au
Sign up to our email alerts: publish.csiro.au/earlyalert

CSIRO Publishing publishes and distributes scientific, technical
and health science books, magazines and journals from Australia to a
worldwide audience and conducts these activities autonomously from the
research activities of the Commonwealth Scientific and Industrial Research
Organisation (CSIRO). The views expressed in this publication are those
of the author(s) and do not necessarily represent those of, and should not
be attributed to, the publisher or CSIRO. The copyright owner shall not be
liable for technical or other errors or omissions contained herein. The reader/
user accepts all risks and responsibility for losses, damages, costs and other
consequences resulting directly or indirectly from using this information.

Designed by Bobby Birchall, Bobby&Co.
Reproduction by Saxon Digital Services
Printed by Toppan Leefung Printing Limited, China

Insect font © István Orosz; Back cover *Gauromydas heros,* a mydas fly
© Trustees of the Natural History Museum, London

Contents

Introduction

THEY OUTNUMBER us 200 million to one and long before us humans stamped our mark on this planet, insects had shaped it into the colourful and extraordinary world we all share today. To many of us they are 'creepy crawlies' best left in the undergrowth, but entomologists (researchers who study them) know better – insects are the lifeblood of this planet, encompassing pollination and so much more.

Their presence has shaped our cultures for thousands of years. The larvae of moths, the workers of bees and the defence mechanisms of true bugs have provided the silks, waxes, and dyes with which we have documented our progress through time. They have been a part of our diet for millennia – any readers who are currently on a palaeo diet, go ahead and munch on some migratory grasshoppers. We can dwell on the negatives – their perceived filth, or their key role in disease transmission – but these tiny but mighty creatures have led to extraordinary discoveries, from robotics to genetics to forensics, transforming our knowledge of agriculture, evolution, medicine,

A common garden visitor in the UK announcing the start of spring –
Bombylius major Linneaus 1758, just one of the many millions of species of animals described.

aerospace, artificial intelligence, biodiversity and ourselves. Just one species of fly has helped us both look after ourselves on Earth and now in space. Several species of beetles are helping us live in the most hostile of habitats. And thousands of cockroaches are telling us amazing things about the physiology of all animals.

Our knowledge about insects is as proportionally as small as they are to us. We don't even know how many there are out there. The latest count of mammal species as of September 2022 was 6,495. That may seem like an impressive figure, but it really isn't in comparison to the number of insects. Around one million insect species have been described to date but figures for the total number of species is vast – figures range from five million to 2.2 billion. That range is enormous, but even if we take the most conservative of estimates – that's still a lot of little things.

To understand our environment, we started naming it. And we all know what humans are like at naming things – those in the UK may smile thinking about our recent history of *Boaty McBoatface* for a research vessel (this was deemed not appropriate, and the boat was named *RSS Sir David Attenborough* though the name lives on in the public consciousness as an autonomous underwater vehicle). But names have been problematic since we started using them.

Naming is not new – the scientific field that deals with nomenclature (the assigning of names) is called taxonomy from the Greek meaning 'arrangement method', and it dates back three millennia. Oddly, the mythical Chinese Emperor Shennong (or Shen Nung), a deity in both Chinese and Vietnamese folk religion who was credited with the invention of agriculture (inventing implements such as the hoe and the plough, as well as the use of boiled horse urine for preserving seeds) and the use of plants for medicine, is also given credit for kick starting us off in organizing the life around us around 3000 BC. From 206 BC this knowledge was finally transcribed into a series of books – the *Shennong Bencaojing*. The first book included 120 drugs harmless to humans, the second book 120 substances that are beneficial, whilst the last book identified 125 drugs that weren't. I feel it may have been better the other way round!

From China across to India, and the naming of animals had kicked off before these publications around 400 BC with Indian physician and

Shennong with his plants – painting by Guo Xu in 1503.

scholar Charaka who was happily classifying 'species'. He had divided up the animal kingdom into four groups: those born from the uterus i.e. humans and other placental mammals; those born of an egg; those born of moisture and heat e.g. worms and mosquitoes; and those born of vegetable matter. This may seem a little amusing to us now, but this was two and a half thousand years ago! Just a hundred years later another Hindu scholar, Prasastapada, developed a new classification with two main categories: asexual and sexual, with the latter split between sexual and oviparous or egg laying. Specifically with the insects, for that is where it really gets all messed up and complicated, Sustuha, the Hindu scholar from 100 AD split insects into three: krimi – arising from moisture, faeces, and from decomposing dead bodies; kitas – venomous insects and also large scorpions; and pipilicas – ants, mosquitoes, gnats, and similar insects. It was to be a long time till we understood how insects were 'born' (and it is covered in this book).

But as these works and traditions were not known to the western world till the Middle Ages and so we give much more credit to the Ancient Greeks and the Romans. The Greek Aristotle (384–322 BC) has been described as the first to classify all living things, an incredibly bold and inaccurate statement considering how little we still know! Plinius (23–79 AD) or Pliny the Elder, an army man but also an author and naturalist described many species in his *Naturalis Historia* (it was after all 37 volumes), a publication now described as the first scientific encyclopaedia.

By the 1500s, many different individuals from across the globe were getting in on the act – so many names, but very little conformity. Many names were polynomic which although descriptive weren't necessarily helpful, 'Caryophyllum saxatilis folis gramineus umbellatis corymbis' meaning 'Caryophyllum growing on rocks, having grass like leaves with umbellate corymbose inflorescence'. Can you imagine if names were to grow the more you learnt about their ecology – 'Erica daughter of McAlister currently living in South London, having slightly unkept hair and a poor but keen singing voice? Adrian son of Washbournes currently in the wilds of southern England, having less hair than the aforementioned species but probably a better voice.' Things get messy and complicated and that's without the obvious problem of using many different languages. Roll on to 1623 and to a Swiss botanist (1560–1624) Gaspard (or Caspar) Bauhin and the publication of his book *Pinax Theatri Botanici*, which includes thousands of botanical names that were classified in a fashion not dissimilar to the later but more comprehensive system produced by Carl Linnaeus (1707–1778) (also known as Carl von Linne, Carolus Linnæus and Carolus a Linne – absolutely no irony in the multiple names).

Linnaeus brought the scientific community together with his landmark binomial naming system. The first edition of *Systema Naturae* was published in 1736, but it wasn't until 1758 and the tenth edition that zoological nomenclature was born. Linnaeus set about creating stability out of chaos – the wanderlust that drove individuals to explore our world created a hodgepodge of knowledge. Starting in his native Sweden, Linnaeus also travelled through Lapland, France, and the UK, naming species as he went. And it is in the UK that his famous collection resides, acquired from his widow Sara Lisa in 1784 by the British botanist Sir James Edward Smith (1759–1828) on the recommendation of another British scientist Sir Joseph Banks (1743–1820). For a mere 1,000 guineas (it's gone up a bit) 14,000 specimens together with books and manuscripts were floated across the North Sea to be housed in a

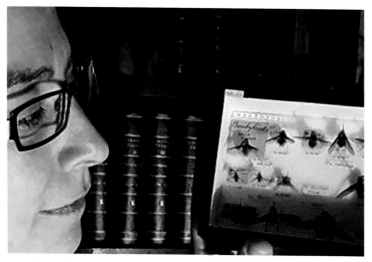

A very happy fly specialist looking at the type specimens of bee flies (Bombyliidae) described by Linnaeus in 1758 and held in the vault of the Linnean Society.

bombproof vault underneath the streets of Piccadilly in London, UK – the home of the Linnean Society of London, which Smith had founded. This society may not be a household name to many, not even to many scientists, but it was pivotal for those interested in how we categorized our world. The species held in the vault are the name-bearing specimens that all taxonomists refer to when either checking on existing species or creating new ones, which we refer to as types.

Binomial species' names, originally in Latin but now very much a mixture of many different languages (that have been Latinized), are followed by the surname of the person who described them and the year that the name was accepted. Just like life, taxonomy is always changing and often the species are found to no longer belong to the original genus from which they were placed, and so they are moved – and this is indicated by a bracket around the name and date. To confuse things a little zoologists and botanists do this differently. Botany has a list of standardized abbreviations for authors and if the species has changed genus – the author who did this is written. So for example, the pedunculate oak, *Quercus robur* L. with the L. being Linnaeus, whilst the sessile oak is *Quercus petraea* (Matt) Liebl. as Mattuschka described it originally and Lieblein moved it to the *Quercus* genus. To ensure that us taxonomists still maintain order in our naming a set of rules were developed for zoologists (*The International Code for Zoological Nomenclature*) and botanists (*The International Code for Botanical Nomenclature*), and alongside this we formalized the principle of typification, i.e. the rules of applying names. Now we recognize primary types (those with name-bearing status) and secondary types (those without but for several reasons are valuable e.g. maybe caught at the same time or place). Throughout this book, the names and dates

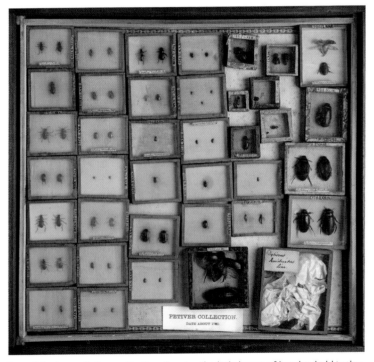

Part of the James Petiver collection showing the little boxes of beetles, held in the Natural History Museum, London.

(when zoological) are next to the scientific names to give readers an indication of just how many people are working tirelessly, and often for no money, describing new species – the efforts of so many hidden in the archives of time.

And there is a much larger beast of a collection just down the road from this illustrious vault, housed in what is the Natural History Museum, London. Within its hallowed walls there are over 80 million specimens of which there are approximately 34 million insects, ranging in size from the smallest wasps, *Dicopomorpha echmepterygis* Mockford, 1997, whose size of 0.127 mm is smaller than a full stop, to one of the longest *Phobaeticus chani* Bragg, 2008 or Chan's megastick measuring 567 mm (22¼ in). This collection is notable for many things including Joseph Banks's collection of specimens that has specimens from Captain Cook's circumnavigation of the globe on HMS *Endeavour* or the even older James Petiver (1865–1718) insect collection where the insects were housed in small boxes or between mica sheets! But more importantly it contains one of the largest collections of type material – approximately 250,000 primary type specimens (i.e. not including secondary types).

But insects aren't just a name or a function – they are a form. And what a marvellous variety of form they exhibit. It is this variety that helps us describe one from another, and it is these changes in form that have enabled them to get their tarsi into nearly every ecological niche on the planet. And it is this extraordinary variety that is at the heart of this book. We have looked at these small creatures for so long, describing what we see, but it's only recently that we have been asking different questions about them. Figures from the past that we know much about, as well as the ones whose work has been hidden in the archives of journals their names long forgotten to all but a few, have got us thinking about what really is happening in this minuscule world. But it is the tenacity of all these men and women that have led to some of the most amazing discoveries, often resulting in new bioinspired technology that can help us survive on this planet alongside our tiny brethren. Turn the pages of this book to read about individual journeys of discovery, the insects that they championed and from which have arisen some truly innovative developments.

Gallus domesticus.
Colombo, Ceylon. 1907.
Albert J. Chalmers. Coll.

CHAPTER ONE

Jumping Jack flash

Adam
Had'em

Fleas, Ogden Nash

I N ALL OF NATURE, there is hardly a more impressive athlete than the tiny flea. These amazing, wingless creatures are not only able to jump distances of more than 60 times their body length, but they also make these jumps repeatedly, apparently without tiring. Previously considered the irritating bedfellows of everyone from kings and queens to the poorest in society, it has taken three centuries for scientists to reveal this mini marvel's abilities. The most meticulous of research about the phenomenal jump of a flea has gone on to inspire a wealth of applications, from innovations in human medicine to the incredible engineering of micro-robots.

Our knowledge of the flea has come from many and varied sources. On my desk at work for instance, at the Natural History Museum, London. I have the assortment of usual things that all folks have: a phone, a laptop, a microscope... and a chicken's head – collected by whom is sadly not known. I'm not uncivilized, the head is safely bathed in ethanol, in a beautiful sealed glass container from when it was collected in Sri Lanka in 1907. And it is not so much the head

'In the eye of the beholder' – a chicken's head with sticktight fleas embedded in the eyelid.

I'm interested in, rather what is buried within. The chicken's eyelid, is riddled with fleas, dozens of tiny, fascinating, little fleas.

They belong to the species *Echidnophaga gallinacea* (Westwood, 1875), also known as the sticktight flea since it does exactly that. This species is rare among fleas. While many jump around on their host to avoid being detected or removed, these ones live around the eyes where preening is less than efficient. Living on the eyelids of chickens, other birds, dogs and even the odd human, the adult fleas are adapted for hanging on and feeding rather than evading a sharp beak or a clawed foot. With a fixed head and a formidable mouthpart, they remain embedded in the host's eyelids, and if they are female, this set up is permanent. She stays put, shooting her eggs out with considerable force to ensure they fall clear of her host. Interesting as all this is, these fleas do not jump as adults, and only really perform little 'hops' whilst juveniles, so reveal little about the athletic abilities of their relatives. I just like them.

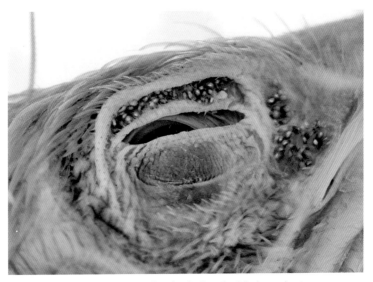

Snuggle up little ones – dozens of sticktight fleas huddled together in a blink of an eye.

Most of us have come across fleas in our lives – thanks to our love of fluffy household occupants. One of the reasons I ended up as an entomologist, as well as working hard and doing my homework, was that as a child I was given a microscope. It was an old teaching microscope that was being thrown out by a school and thankfully it ended up with me. As part of my very outdoors upbringing, I had a pet goat, but we also had chickens, ducks, geese, rabbits, the odd guinea pig and of course the usual collection of dogs and cats. The cats were able to roam around and often they would bring me gifts, which led to my fascination with maggots – but that is a later story. The cats themselves though would also bring in fleas. I'd try and catch the fleas, and then flounder around trying to recatch them as they jumped out of my grasp. Fleas, I discovered, were a marvel. I'd swear that I had one between my fingers only to open them and find nothing. If I was successful though I'd place them into a clear box and watch them under the microscope. Lots of giggles and muffled whooping as they kept jumping up, straight at the lens, causing me no end of amusement.

These insects don't fly to escape, they mostly jump. They are descended from winged ancestors, being closely related to flies (Diptera) and scorpionflies (Mecoptera), and the relationship between these three orders has been debated for a long time. Recent research by Erik Tihelka and colleagues at the University of Bristol, UK, now points to them having descended directly from scorpionflies. They started as plant-suckers, but then (this was around 290–165 million years ago) developed a taste for more nutritious hosts – mammals, and from them onto birds. With these changes came an adaptation for explosive and rapid jumping over flight.

Robert Hooke (1635–1703) wrote (and drew) fleas in his wonderful 1665 book *Micrographia or Some Physiological Descriptions of Minute Bodies Made by Magnifying Glasses with Observations and Inquiries* – a title as big as the organisms it related to were small. Through his close observations he noticed something rather special about their legs, '... that he can, as 'twere, fold them short one within another, and suddenly stretch, or spring them *Schem*. 34 out to their whole length, that is, of the fore-leggs, the part A, of the 34. *Scheme*, lies within B, and B within

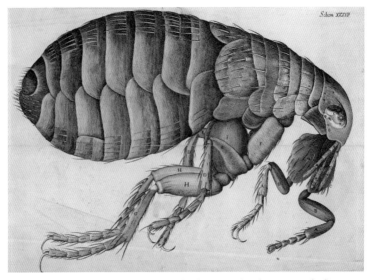

Robert Hooke's flea, annotated with letters that he uses in his description of a flea jump.

C, parallel to, or side by side each other; but the parts of the two next, lie quite contrary, that is, D without E, and E without F, but parallel also; but the parts of the hinder leggs, G, H and I, bend one within another, like the parts of a double jointed Ruler, or like the foot, legg and thigh of a man; these six leggs he clitches up altogether, and when he leaps, springs them all out, and thereby exerts his whole strength at once.'

This detailed observation of their fantastically springy legs is a well-shared nature fact, hundreds of years on – these species can jump! I was often told as a child the story of them being able to jump over St Paul's Cathedral, UK, if they were but human size. But is that true? Perhaps, if we disregard all principles of biology and physics (the whole problem with enlarging insects and how it would move/support itself etcetera), and just simply 'scaled it up' then, taking my cat's fleas *Ctenocephalides felis* (Bouché, 1835) as an example, here's the answer – normally this 1.5 mm (0.059 in) flea can jump 50 times its height, the equivalent of me jumping 80 metres or 262 feet, and so yes – ignoring science – a flea could jump over St Paul's if it was the size of a human.

But how do they do it? Well much of our insight begins with the esteemed British naturalist, the late Dame Miriam Rothschild (1908–2005), and what turned out to be her life-long fascination with these acrobatic creatures. She famously described them as 'insects that fly with their legs'. Wings, as she rightly pointed out, are no good to you if you live in fur. The name Rothschild was synonymous with banking, and Miriam's father, Nathaniel Charles Rothschild (1877–1923), known as Charles, was heavily involved in the family business. He was reported to have never missed a day's work in his life! But his heart lay in fleas, and during his short life he amassed a collection of more than 260,000 of them, containing 925 types, those revered specimens that are the first of their kind to be found. All his fleas are at the Natural History Museum, London given as a deed of gift in 1913, but not registered as Museum property until Charles's death in 1923. In case you don't realize how important his bequest was, the collection contains just over 70% of all described species and subspecies of flea. Known as the Rothschild Collection, it is one of the few still in its original beautiful cabinets and I, as the curator responsible for them, love them.

The Rothschild Collection. The cabinets, the monstrosities and freaks.

A bride and groom. Dressed fleas from Mexico, 1905, the outfits adorned with flea heads.

Charles travelled extensively during his life, and the collection reflects that – the geographic range is worldwide. It contains one or two oddities, too. For example go to the Natural History Museum at Tring or drop by my desk, and you will see some unusual fleas from Mexico, collected in 1905. These *Pulgas Vestidas* are unusual in that they appear to be wearing clothes (the Spanish name means dressed fleas), a craft that may have originated from convents in Guanajuato, spreading out amongst local villages and becoming something akin to a tourist trinket. Of course the fleas weren't actually dressed, they were used as the tiny head on a body of a tiny model. Such was Charles' obsession with fleas that in January 1914 the *Daily Mail* published a piece claiming 'Mr Charles de Rothschild, of London ... paid £1,000 for a specimen of a rare variety of flea, one of the kind which is occasionally found in the skin of a sea otter'. Charles responded by saying he had 'never paid any such sum for a flea of any description'. He did, however, collect and describe more than 500 species new to science, and set new standards of definition and accuracy in flea identification.

Charles married his cousin Rozsika von Wertheimstein (1870– 1940), whom he met during a butterfly excursion in the Carpathian Mountains in central and eastern Europe, and the pair raised their four children back at Ashton Wold in Northamptonshire, UK. Miriam Louisa Rothschild was their first, and from an early age became interested in natural history – hard not to really, when your father was a pioneer for nature conservation. I have collected flies at Wicken Fen, Cambridgeshire, UK, which, mostly thanks to Charles, was the UK's first nature reserve, now looked after by the National Trust. Not only that, in 1912, thanks to his concern over the state of the British countryside, he organized a special meeting at the Natural History Museum, London from which he formed the Society for the Promotion of Nature Reserves (SPNR), which gained its first Royal Charter by George V in 1916. You might know it better as The Royal Society of Wildlife Trusts or simply The Wildlife Trusts – who now manage 2,300 nature reserves across the entire UK.

Although Charles had been educated at Harrow School in London, formal education, he believed, 'crippled' young, especially bright girls'

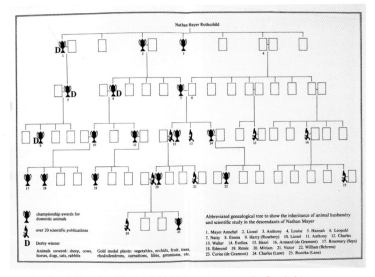

From her book *Dear Lord Rothschild*, Miriam charts out the family history with both sport and scientific successes.

minds and so for his daughter Miriam the garden became her classroom, her home was her laboratory, and her books, the people and the world around her became her teachers. Miriam gained no formal training but amassed eight honory doctorates and became a leading authority on all things flea. As she said in a rare BBC interview back in 1995, 'They're rather sweet aren't they. It isn't everyone that has a great love of fleas, but I have. These fleas can jump 30,000 times without stopping, which is really rather a lot.' You can hear in her tone how much affection she held for these animals. She published more than 300 scientific papers, became the first female committee member for the National Trust as well as the first female trustee of the British Museum of Natural History, now the Natural History Museum. She joined her brother as a Fellow of the Royal Society, and to date they are the only siblings to have achieved this. Quite simply a true trail blazer.

A former Keeper of Entomology at the Natural History Museum, Dr Richard Lane, fondly remembers meeting her, and how her first real claim to fame was working on her father's great flea collection and writing its extensive catalogue. Lane stresses this isn't just a list of names but a compendium of infomation associated with their biology, history, distribution and so on. Her visits were frequent to the Museum, he says, dressed in moon boots, and great flowing gowns, often accompanied at Christmas with pheasants for friends and colleagues (including Dr Lane).

And remember those beautiful cabinets that her father's collection resides in? Well alongside them are many more from Miriam's work – some are labelled ova, larva and pupa and relate to her serious studies researching the life cycles of fleas, while others have less scientific labels. One is 'Odds and Ends', because who doesn't have specimens that don't fit in? And my personal favourite, 'Monstrosities', which houses specimens with slightly larger or wider features than the average. One slide, housing a flea described as having especially large genitalia, has the word 'freak' scribbled on it.

By the mid 1960s, Miriam had discovered, among other things, that the life cycle of the rabbit flea *Spilopsyllus cuniculi* (Dale, 1878), the vector of the deadly myxomatosis disease, was controlled by the sex hormone cycle of its host (published in *Nature*, no less). The flea in this instance was timing its reproductive cycle with when the rabbit's gave birth – baby rabbits meant many more hosts for baby fleas. But what Miriam was most interested in was, of course, their jump. She described it as one of the most extraordinary observations of entomological agility, 'they simply take off and disappear'. Most flea legs

Xenopsylla philoxera (Hopkins, 1949), a flea collected from the white-toothed or musk shrew *Crocidura deserti* (Schwann, 1906).

average just 3 mm (0.12 in) and so, without much length, they have little time to generate energy to push off the ground. So how do these tiny insects achieve these amazing feats? Miriam set about building on Hooke's earlier observations with her usual tenacity, and conducted many intricate dissections of the flea's anatomy, producing her own beautifully detailed images of their joints and tissues. Later, in 1986, alongside Prof. Yosef Schlein and Prof. Ito Shosumo, she would publish *A Colour Atlas of Insect Tissues via the Flea*, which is not only a great title but features a flea's vagina as its front cover. It was a bold statement and precursor to an extravagant compendium full of preparations of all flea parts including their leg muscles. However, the flea's jumping, she claimed, couldn't just be down to muscles. Their jumps were known to be quick and energetic, so the power requirements were way beyond what a muscle alone could produce by a direct contraction. For one thing, even the fastest possible single muscle contraction would fail to move the legs rapidly enough. Moreover, muscle efficiency tends to decrease with temperature, yet the flea seemed indifferent to cold. Even when the temperature dropped close to freezing, its jumping ability was largely unimpaired.

Fleas have adapted the powerful muscles used in other insects for flight, for jumping. They are contained in the middle section of the body, known as the thorax. In a 1975 paper by Miriam Rothschild and Yosef Schlein, they describe the jumping mechanism of the oriental rat flea *Xenopyslla cheopis* (Rothschild, 1903), (a species incriminated as one of the major vectors of the plague and described by her father) using two beautiful line diagrams (see pp. 26 and 27). The first illustration is a parasagittal section of the metathorax (the final section of the thorax) and hind coxa (the segment that joins the legs to the thorax); and the second is a transverse section though the metathorax (middle part) and hind coxa again, highlighting the tendons and the muscles. In the paper they note the massive coxa, but they also comment on large air sacs further down the leg, which they presume aids buoyancy when the 'insect is airbourne'. These creatures have honed their bodies completely!

What these intricate drawings reveal is, first off, how large the muscles are and, secondly, a highly modified part of the internal part of the exoskeleton called a pleural arch, within which an elastic protein called resilin is found. It is this rubber-like substance that holds the secret behind the power of the jump. Dr Greg Sutton of Lincoln University, UK, a specialist in the biomechanics of insects, describes resilin – in which it stores energy before a sudden release – as the secret behind the flea's powerful jump. Alongside the pleural arch itself are internal thickened ridges as well as hooks, pegs and catches that enable the animal to clamp the thoracic body plates together to provide more support for the muscles to act against. It creates, in effect, a highly efficient jumping machine. First the flea locks its legs in place, which packs energy into the pad of resilin. With the energy in this spring now stored, the flea can release the spring, shooting its legs out much more quickly than muscles could ever do alone. It's mechanically identical to when we use an archery bow to fire an arrow. Our muscles can't generate enough energy to fire an arrow very far. Instead, they generate the energy by stretching the elastic of the bow. When we release the bow it returns the energy stored in the bow to the arrow much more quickly.

It turns out that the force generated – the g-force – by a jumping flea is an incredibly impressive $140g$ (140 times the force of gravity). To get an idea of the power, the force of gravity when you sit or lie down is considered $1g$, but an increase even up to $4g$ will cause you problems. Chief medical reporter at CNN Dr Sanjay Gupta did some research before his own personal experience of being taken on a high-speed ride of his life, and detailed what happens in his blog, 'From a medical standpoint, at 4 Gs you will start to lose colour vision, which is why it is called 'graying out' — 4.5 Gs and you may lose vision all together. Higher Gs and your lungs start to collapse, your oesophagus stretches, your stomach drops and blood pools significantly in your legs.'

Astronauts reach $3g$ while some fighter pilots reach $9g$. Exposure over $16g$ is deadly for our soft-bodied species. It appears the smaller you are the more g you can handle. One species of tardigrade, *Hypsibius dujardini* (Doyère, 1840), has been shown to withstand $16,000g$. Going back to the flea, the force of its acceleration, according to Miriam, is 20 times that of a space rocket re-entering Earth's atmosphere.

Resilin was first discovered in the wing hinges of locusts by Danish zoologist Torkel Weis-Fogh (1922–1975). Ironically for this story, he is

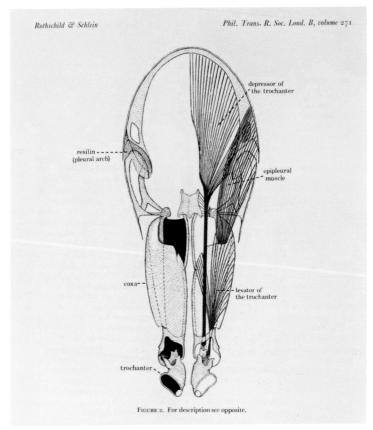

Rothschild & Schlein　　　　　　　　*Phil. Trans. R. Soc. Lond. B, volume* 271

FIGURE 2. For description see opposite.

Sections through the oriental rat flea, *Xenopsylla cheopis,* to show main jumping muscles (above and opposite). The tendon of the trochanter (second leg segment) and the trochanteral depressor are coloured red. The levator, epipleural and pleurocoxal muscles are coloured blue. (See p. 24).

more famously remembered for his work on insect flight. The zoologist Henry Bennet-Clark, who was to become a key player in the jumping flea's story, and now an emeritus professor at the University of Oxford, UK, described with great excitement hearing Weis-Fogh reveal his discovery of resilin in a seminar in the late 1950s. During a stint at the University

Rothschild & Schlein — Phil. Trans. R. Soc. Lond. B, volume 271

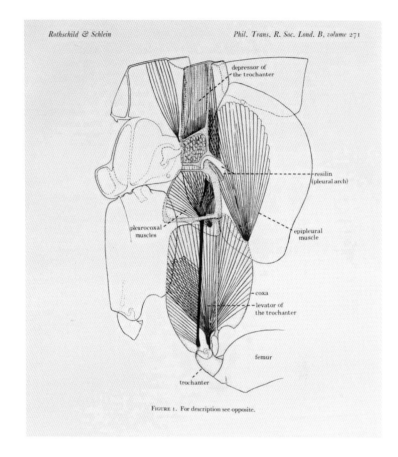

FIGURE 1. For description see opposite.

in Cambridge, UK, Weis-Fogh had discovered that within the insect cuticle, a major part of the outer layer of the insect's exoskeleton, was this rubber-like protein. He was able to show that this protein could withstand hundreds of millions of extensions and contractions for many weeks, without losing its form – it basically sprung back into shape everytime.

Bennet-Clark describes resilin as an 'elastic spring antagonist to muscle', and in 1967 co-authored a paper describing the spring and how it powered the high-speed catapult used for flea jumping. More recently, biomechanic Sutton has described this spring as a sandwich (albeit a blended sandwich) consisting of resilin that the insects manipulate, aligned with a saccharide called chitin that is very, very tough (I am presuming this is a very old sandwich). Chitin is technically an aminopolysaccharide polymer and as polysaccharides go its abundance on this planet comes only second to cellulose, which is found in the cell walls of plants and algae. Although its structure is similiar to cellulose, its form and function is more like keratin – a protein very familiar to us in the form of hair, nails, claws and hooves. Resilin can be highly flexible and deforms well. It's this sandwich layering of chitin and resilin that also resembles the design of an archer's composite bow from hundreds of years ago, made of horn and wood, and as such, both the bow and the protein are able to store a lot of energy. So this interface between the protein and the saccharide enables the spring to be loaded and held in position for a long time. The spring can then recoil with almost 100% of the energy that was put in. It's these unique properties, of possibly the most elastic substance we know of in the natural world, that have begun to attract attention far beyond the realm of entomologists.

Kristi Kiick is a professor of biomedical engineering at the University of Delaware, USA, and she and her research group have had a long-term interest, over the past 15 years or so, in trying to apply resilin in different biomedical applications. Inspired by its resilience, the Kiick laboratory has been synthsizing this amazing rubber protein. Their aim is to model and mimic resilin's ability to stretch and recover, and so to regenerate damaged human body parts that under normal conditions undergo repetitive strain at high frequencies – such as the vocal cords in the throat.

The Kiick lab team has been engineering resilin-like polypeptides (RLPs) that possess all those favorable attributes of native resilin. They have gone on to develop RLP-based hydrogels, that can form well-defined shapes and that also have good cell-adhesive properties. It's physical hydrogels such as these that are now becoming important for biomedical applications since, not only are they compatible and non-toxic to humans, they also have that crucial reversible elasticity. At a general level, Kiick and her team are developing ways to vary the degree of elasticity that could perfectly match damaged tissue within the body, and to then inject the sythesized resilin protein hydrogel wherever more is needed. More specifically, they've recently shown it works as potential material for cardiovascular tissue engineering. Hybrid heart valves for the future? Here's hoping.

I love the idea that a whole new class of materials has been developed thanks to the discoveries of inquisitive entomologists. Miriam had answered one question about the fleas jump but another still remained unanswered – since fleas and their complicated legs always jump off at an angle, how do they launch? How is the force of the spring transferred through the leg to the ground? The answer as to how fleas harnessed this explosive energy was thought to be the key to understanding how the fleas control speed and direction of the jump – and it was this that had preoccupied flea fanatics for decades.

By the early 1970s, two competing hypotheses had emerged. Rothschild argued that fleas planted a knee-like joint called the trochanter to jump. Bennet-Clark, on the other hand, thought the fleas pushed off from the footlike segments at the end of the legs, called the tarsi. They both drew upon the still very infant technique of high-speed photography, building upon techniques first used by Eadweard James Muybridge (1830–1904), the eccentric and infamous landscape and motion photographer who had shown how to freeze time in photographs and reanimate these images both for scientific analysis and entertainment. Born Edward Muggeridge, in Kingston-upon-Thames, UK, he lived a dramatic life following emigration to the USA. He travelled extensively through Central America, sustaining a serious head injury in a stagecoach crash, and ending up murdering his wife's lover AND getting away with it. Muybridge's instantaneous photography had made motion visible, and from the late 1870s he took thousands of images studying human and animal movement. It was in California after a rumoured $25,000 bet that his fame was secured, capturing a horse's movement on film to decipher whether all its legs left the ground or not when in full gallop. It involved the ambitious Leland Stanford, founder of Stanford University, USA and owner of champion racehorses with an obsession to breed yet faster champions. To settle a bet he's said to have made with the *San Francisco Chronicle*, Stanford commissioned

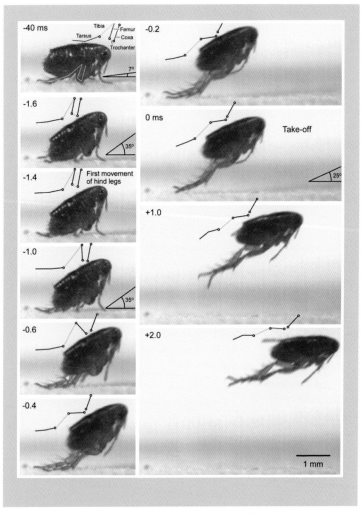

Jump. Jump. Go ahead jump. The images from the work of Dr Sutton and Prof. Burrows on flea jumping that established whether the force for jumping went through the knee-like trochanter or the tarsus (foot).

Muybridge, armed with rows of cameras and trip wires, to capture split-second movements from a horse on his farm in Palo Alto. The horse did indeed lift all four legs off the ground during its stride. However, this was not in the front-and-rear extended posture some had expected, but with all four feet tucked under.

Rothschild and Bennet-Clark set about using film in the same way, to capture the much smaller animal's jump. Not an easy task when you consider it takes a mere one millisecond. Just for scale, the blink of an eye is 100 milliseconds long. The pair separately employed mechanical high-speed cameras, running their kit at incredibly high speed with giant rolls of film and then hoping and praying the flea jumped within just the three seconds-worth of film that a single reel could accommodate. At the time, these high speed films were cutting edge. The push off phase of the jump was maybe one frame, maybe two. Bennet-Clarke had teamed up with the scientific film pioneer Eric Lucey who, in 1966, had been commissioned by the BBC to capture the jump of a flea at thousands of frames per second using a state-of-the-art Fastax high-speed camera. Rothschild's later set-up in her lab in 1972, employed a little pyramid for the fleas to ascend. She describes them as sitting at the top and so enabling the camera operator to focus in detail on the animal. And if you've ever tried to make a flea jump on cue, and in the right direction, you'll know that's hard. If you haven't, then you can well imagine. Despite reels and reels of film the answer was to prove elusive. Both Rothschild and Bennet-Clark agreed on the starting position. But neither had enough data to go deeply into the one to two milliseconds when the force was being generated by the flea, to figure out whether it had gone through the knee-like trochanter or through the tarsus, or foot.

Forty years were to pass before this riddle was solved. And it was Dr Greg Sutton, and his then fellow University of Cambridge colleague Professor Malcolm Burrows, who in 2011 published their paper on the biomechanics of jumping in the hedgehog fleas *Archaeopsylla erinaceid* (Bouché, 1835), supplied from St Tiggywinkle's Wildlife Trust in Aylesbury, UK. They did this in two stages. First they constructed a model to simulate the velocities and accelerations that the two competing hypotheses would have produced, and secondly they undertook their own high-speed photography of these jumping fleas. And cameras by then were a lot better – running film at 10,000 frames per second. Finally they were able to answer the question – they were able so see that the foot was on the ground during launch. The insects were transmitting the force from the spring in the thorax through leg segments acting as levers to push down on the tarsus and launch the tiny animal at speeds as fast as 1.9 m/s. As well as seeing this, they took some scanning electron microscopic images that showed how spines along the tibia and tarsi helped in increasing the surface area through which force could be applied to the ground to propel them.

Propelling away fast by harnessing all these mechanical propterties is one of several important lessons fleas can teach engineers. Getting up off the floor and over obstacles without flying is now fuelling a new generation of jumping micro-robots. Professor Sarah Bergbreiter is an electrical and computation engineer at Carnegie Mellon University, USA. Jumping, she argues, offers the most efficient locomotion on uneven surfaces for millimetre-sized micro-robots, and she is developing autonomous jumping robots, with power and control on board, that are as small as grains of rice. The possibilities of these miniature robots are limitless – providing cheap sensing or surveillance in inaccessible areas,

A branch jumper – one of Bergbreiter's bio-inspired robots.

stealth tracking or helping with search and rescue through rubble after natural disasters.

Like a living flea, the bio-inspired robot uses precisely matched components that work together to enhance performance, store energy required for jumping, and then release it quickly when needed. This was not easy. Bergbreiter's team began by experimenting with materials on larger-sized robots and then scaled them down, using a combination of rigid silicon (the same material used to make integrated circuits) and silicone, the rubber that is very much like resilin. In essence they created a spring-actuated system whose motion is separated into distinct phases. An actuator such as a motor (equivalent to a muscle) stores energy in a soft rubber spring that is then held in place by a contact latch. When the latch is removed (also by an actuator) this energy is quickly released. They were able to get jumps of more than 30 cm (11¾ in) from a micro-robot just a few milimetres wide! They are now looking to improve control of these prototypes, such as varying the speed of latch release, which could provide a means for these robots to jump varying distances. If the latch motor is fast enough and strong enough, the robot could cover a full range of jump distances from zero to its maximum jump height or distance defined by the energy stored in its spring. Again it's the combination of rigid and soft that has enabled this – much like the body of a flea.

Despite all we know, the tiny flea has yet to reveal all its intricate secrets. No one yet knows how it locks its spring in place and then releases it. And no one knows how a flea snaps its two hindmost legs at exactly the same time. If fleas weren't so precise, they'd spin wildly off course. So the story of the flea remains a good one, and it will continue to enlighten us. It is a story of an animal that has evolved over millennia for a body shape and design that is ideally suited to life on its host. The larger and more active the host, the greater the flea's jumping ability. And while some may consider this humble creature to be a pest, as we learn more and more, we can also consider it an inspiration. Once again we are reminded that insects were often doing things before us and, most of the time, way better.

CHAPTER TWO

Mighty mouthparts

*Beside me on the rail, the sphinx moth races its engines for
take-off like a jet on a runway. I could see its brown body
vibrate and its red-and-black wings tremble*

Annie Dillard

WITHIN THE Natural History Museum collections in London
there are many Lepidoptera – the insect order that comprises
the butterflies and moths. And I mean many. There are more than
13 million of them, housed over four floors. This is all the more
remarkable a figure when you find out how small some of them are. The
micromoths are incredibly small (the clue is in the name) where one of
the smallest species, *Stegmella maya* (Stonis et al, 2013), from Mexico
has a wingspan only measuring 2.8 mm (0.11 in). It's a mega-rich order,
with more than 180,000 species (many of which are a lot larger) in
126 families, described in thanks in no small part to the intrigue of
the Victorian entomologists who were seduced by their many pretty
patterns and ease of identification. But scratch the surface, or should
I say scales, and there is a whole lot more to these creatures. And it
is indeed one of the Lepidopteran families that we turn to next for
inspiration, the Sphingidae or hawkmoths.

I personally think these moths look like little stealth bombers, or
jets as the twentieth-century American author Annie Dillard wrote.
But my colleague hawkmoth expert Dr Ian Kitching calls them the

The elephant hawkmoth *Deilephila elpenor* (Linnaeus, 1758).

sports cars of the insect world, for they are the sleekest and arguably the most glamorous of all the moths. Kitching himself has researched their evolutionary trade-off between body size and their distinctive wing shape. He has worked on hawkmoths for 35 years, has described 47 species and subspecies, and written two comprehensive checklists of them, both from the collection and out in the field – so it's fair to say he knows them. 'It's a beautiful thing to watch these creatures as they hover over the seductive nectaries of plants', he muses, 'uncurling and teasing their slender thread-like mouthparts deep into the flower, extracting the nectar as quickly as possible, before moving onto the next with equal speed and precision'. And how do we study such amazing species? Well, when I find myself on a tropical field trip, as is my way, at night I hang up a collecting sheet (a simple bed sheet to normal folks) and switch on a light, then pour myself a glass of wine and wait for them to come. Indeed, moth hunters are blessed with the fact that the majority of our species are nocturnal and a light source and a sheet are all you need to find them.

There are more than 1,700 species of hawkmoths globally. These excellent fliers are very fast and acrobatic. They can also hover, which is when most people spot them, even if they don't at first know what they are. The hummingbird hawkmoth *Macroglossum stellatarum* (Linnaeus, 1758), not to be confused with the hummingbird moth of North America, is a species found across Europe and Asia, and yes is named after its more famous avian doppelganger. Like the bird, this species hovers near plants probing its extremely long proboscis, the elongated mouthparts, down the necks of long-tubed flowers to gain access to their deep nectar reward. Nectar is the bribe plants use to encourage animals to transfer pollen from the stamens of one plant to the receptive stigma of another.

Many flowers seduce animals in this way, and one family in particular – the orchids – has developed an intimate liking for these nocturnal visitors. The orchids, or Orchidaceae, are the plants, like the Lepidoptera, that create the biggest buzz for lovers of natural history. There are around 26,000 species across all continents except Antarctica. That is a lot of species, and they comprise 8% of all described

Colleague Alessandro Guisti collecting moths from a light trap in Borneo.

vascular plants to date, that is those with specialized water and nutrient conducting tissues which are the majority of plants. And we have been fascinated by orchids for a long time. The Chinese philosopher and teacher Confucius (551–479 BC) penned poems about their wonderful smell and many more have been treasured for their strange morphology. And I do mean strange. Check out the naked man orchid *Orchis italica Poiret,* 1799 or the angel orchid *Coelogyne cristata* Lindley, 1824 if you don't believe me. You'll see why the Victorians strongly associated them with seduction. In the western hemisphere, the oldest orchid reference found is by Theophrastus (*c.*371–*c.*287 BC), a pupil of Aristotle, and a scholar considered by many as the father of botany. His book *Enquiry*

The long mouthparts, and hovering posture of the hummingbird hawkmoth, *Macroglossum stellatarum.*

into Plants, contains many original ideas as well as myths about an orchid's suggestive powers.

British naturalist Charles Darwin (1809–1882), known to most of us, also had a love for orchids. He was lured to these flowers like a moth to, well, a flower, in this case. We don't always think of him as a botanist but in 1862, just a couple of years after publishing his sensational *On the Origin of Species*, he produced a rather technical tome on these highly prized plants. Professor Jim Endersby, a science historian at the University of Sussex, UK, and author of the 2016 book *Orchid: A Cultural History*, believes that Darwin's interest may have begun because indigenous orchids grew wild at Down House in Kent, where he lived. Just like Gregor Mendel and his peas, Robert Hooke and his fleas, Darwin drew inspiration from the environment around him. And in observing his surroundings he tried to make sense of the natural world, and in particular, what was driving evolution.

Darwin is credited with kick starting 'modern' biology with his work on evolution by natural selection. It's a process whereby a small change in an organism's genes brought about by a mutation can sometimes result in an advantageous change in their phenotype – their physical appearance – thereby helping them exploit new niches or mate with fitter (in the scientific term) partners. In Darwin's day of course no one understood this mechanism of inheritance, and so it was extraordinary that he was to get so many of the ideas about evolution right. And he was always on the look out to find new and novel examples to prove that his theory was more than just a theory. This is where the orchids come in, in particular their relationship with hawkmoths.

Despite the long-winded title *Various Contrivances by which British and Foreign Orchids are Fertilized by Insects, and the Good Effects of Intercrossing* – his book on orchids contains some neat and extraordinary research. It is full of passages of almost childlike wonder, as he encourages the reader to think themselves into the role of an orchid in an attempt to understand the way pollination could be effectively carried out, playfully seeing the world from a non-human perspective. Even today, in adapting Darwin's fun approach to observing the dextrous pollination by hawkmoths, the behaviour between flower and moth is a source of constant intrigue. Dr Kitching at the Natural History Museum has often described how difficult it is to re-enact the way hawkmoths insert their absurdly long mouthparts into the nectaries of buddleia plants, *Buddleja*. Despite their mouthparts being as thin and as long as a bristle, they hit their target not just once, but repeatedly on the same plant, and with repeated precision. Kitching himself has tried with a replica and failed.

The orchid book was a modest success, as Darwin tried to win over his many sceptics by showing evolution at work, arguing that the way a moth's mouthparts are so precisely designed to fit inside an orchid must be proof they evolved together. Back in 1859 with *On the Origin of Species*, there were many who doubted Darwin's explanation for the

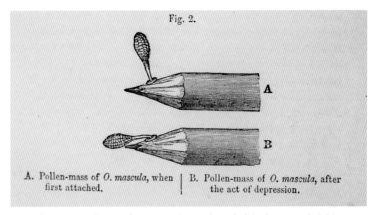

Fig. 2.

A. Pollen-mass of *O. mascula*, when first attached. | B. Pollen-mass of *O. mascula*, after the act of depression.

The pollinium or pollen sac from an early-purple orchid *Orchis mascula* (L.) L. sticking to a pencil. It took just 30 seconds for the pollen sac to desiccate enough for it to achieve the perfect upright position for the next plant's receptive stigma.

many adaptations seen across the living world. Amusingly for me and many of my colleagues, one of the loudest and most hostile attackers was from the Natural History Museum's founder Sir Richard Owen. Himself, a leading comparative anatomist, Owen was open about his disdain for Darwin, who in turn wrote 'I used to be ashamed of hating him so much but now I will carefully cherish my hatred and contempt to the last day of my life'. Sadly, for Darwin, Owen carried a lot of clout. It was only in the early twentieth century, and the rediscovery of the work on inheritance by Mendel and his peas, that we began to understand the basic concepts of genes and their role in heredity, and that Darwin's theory gained acceptance.

When you open Darwin's *Various Contrivances* book, you will find chapters bursting with information on many different species of orchid. He writes about the *Catasetum* Rich. ex Kunth genus as the 'most remarkable of orchids', because when a foraging insect alights onto the plant's landing pad, a hair-trigger is activated that smashes a sack of pollen known as pollinia (singular pollinium) onto the feeding insect's back. The force of the impact wedges the sack onto the creature's back, with a sticky pad at the base to make sure it stays there. Darwin was obsessive in his study of these plants and their entomological transporters arguing that 'the contrivances by which orchids are fertilised are as varied and almost as perfect as any of the most beautiful adaptations in the animal kingdom'.

Orchids have evolved many extraordinary morphological changes to ensure they entice and secure the correct pollinator. Darwin realized in many cases specific insects seem to specialize in specific orchids, and that many orchids in turn are only pollinated by a specific insect – there is a precise lock and key fit. In the case of the catasetum orchid, for example, the large euglossine or orchid bees were the unique key to their lock. This intimate relationship might seem a risky strategy, putting all your eggs (or grubs) in one basket, but in the short term at least, it ensured that the insect was guaranteed no competition for the nectar and increased the chance of it passing on the pollen to the same species. The orchid, therefore, doesn't waste its pollen on incompatible plants, making the process much more efficient.

Darwin's interest in the insect–flower relationship ratchets up a gear when, in January of 1862, he receives an intriguing and unexpected letter and parcel from the wealthy British orchid collector James Bateman (1811–1897). Bateman was a Lancashire-born, Oxford-educated son of rich industrialists whose money had been made in steam power, coal and iron. Also a wealthy landowner and horticulturalist who nurtured a lifelong passion for rare orchids, Bateman had become famous for his implausibly bulky monument to orchid-mania, the trend that had gripped Europe during the nineteenth century – he created the most expensive and ostentatious book on orchids ever produced. It covered the orchids of Mexico and Guatemala, and even within its own pages is described as 'the librarian's nightmare', where a witty vignette by celebrated caricaturist George Cruikshank shows a group of workers armed with pulleys attempting to lift the heavy tome upright so it could be read.

Bateman financed many expeditions to find and expedite orchids back to the UK, where he both studied them but also distributed them to other interested parties. Darwin's surprise parcel contained specimens from Madagascar and amongst them were several, as Darwin describes, 'astounding' star orchids, with their extraordinarily long, whip-like

The 'librarian's nightmare'. The cartoon by George Cruikshank in James Bateman's book on the orchids of Mexico and Guatemala.

nectaries. The orchids were as rare as hens' teeth then, each with a price tag in today's money of more than £10,000. They were *Angraecum sesquipedale* Thouars, which has several common names today including Darwin's orchid, and although first discovered in Madagascar in 1798 by the French botanist Louis Marie Aubert du Petit-Thouars (a name as long as the flower's spur) it was not described by Thouars until 1822.

For Bateman to send three or four of these highly-prized specimens to Darwin indicates the huge level of respect he must have had for him. Indeed, Darwin seems incredibly pleased with the gift and writes that same month to his friend Joseph Dalton Hooker (1817–1911), at the time assistant director of the Royal Botanic Gardens, Kew. 'Bateman has just sent me a lot of orchids with the *Angraecum sesquipedale*: do you know its marvellous nectary 11½ inches [approx. 29 cm] long, with nectar only at the extremity. What a proboscis the moth that sucks it must have! It is a very pretty case.' The scientific meaning of the name

Orch. Afr. O. 19. bis 17.   67

Dolichangis   Angræcum sesquipedale

Thouars's original drawing of the star orchid *Angraecum sesquipedale* Thouars,
also known as Darwin's orchid.

translates as 'measuring a foot and a half' which reflects the excitement of its discovery if not its actual size.

Several months after receiving Bateman's generous gift, Darwin had become convinced it was evidence to help support his theory of natural selection, since it appeared to explain a mystery – how could these prosperously long-necked flowers be pollinated? 'There must be moths with proboscides capable of extension to a length of between 10–12 inches [25–30 cm]' he suggested. If you imagine the moth's mouthparts varying at random and the depth of the nectary of this species of orchid varying at random, Darwin's evolutionary argument was that moths with the longer mouthparts would be slightly better adapted at getting their nectar from deep nectaries, and transmitting the pollen to other plants of the same species, since these moths would be the only insect that could get at the nectar. Less of the pollen would be wasted on arriving on other flowering species it couldn't pollinate, while the moth would get exclusive access to the nectar. And just through random variation and natural selection, gradually you'd get a specialization developing over many, many generations, ending up with the lock and key fit between one plant and one insect. So, natural selection looks like the 'contrivance' producing these amazing adaptations. 'It appears there has been a race in gaining length between the nectary of the *Angraecum* and the proboscis of certain moths; but the *Angraecum* has triumphed, for it flourishes and abounds in the forests of Madagascar, and still troubles each moth to insert its proboscis as far as possible in order to drain the last drop of nectar.'

Darwin had made two daring predictions from seeing these long-necked orchids. The first was the existence of a yet-to-be discovered moth with an extremely long mouthpart. The second, the idea of a co-evolutionary existence between the moth with its long proboscis and Darwin's orchid. Many entomologists mocked his words, but he was so clear on the logic of the theory and the way that the same insight had led him to understand so many other flower-insect partnerships, that he was convinced he was right.

Five years after Darwin's moth prediction, leading British naturalist Alfred Russel Wallace (1823–1913) added further weight to Darwin's idea. In 1867, he published an article entitled *Creation by Law* in the *Quarterly Journal of Science* in which he not only supported Darwin's moth–orchid hypothesis, but also highlighted that the African sphinx moth *Macrosila morgani* (Walker, 1856), now *Xanthopan morganii*, had a proboscis almost long enough to reach the bottom of the spir of Darwin's orchid. Wallace had measured one in the collection of the Natural History Museum, London and wrote, 'Consequently, the deepest nectaried Orchids and the longest nosed moths would each confer on the other a great advantage in the battle of life'. It was an eloquent way of describing this co-evolutionary process. He continued with '… moths do visit Orchids, do thrust their spiral trunks into the nectaries and do fertilize them by carrying the pollinia of one flower to the stigma of another'. Wallace concludes that the changes in length of the plant aren't brought about by the Creator of the Universe, as the Duke of Argyll, a natural theologian, had proposed in retaliation to the *On the Origin of Species*, but by small changes over time that result in such adaptations. So sure of the theory, Wallace went on to say, 'That such a moth exists in Madagascar may be safely predicted; and naturalists who visit that island should search for it with as much confidence as astronomers searched for the planet Neptune – and they will be equally successful!'

It was 41 years after Darwin made his prediction that a moth potentially capable of pollinating this orchid was discovered and named. Sadly, this was also 20 years after his death.

For the next part of the story, we join the Rothschild family, whom we met in the previous chapter with Miriam and her work on fleas. But it is to her uncle Lionel Walter Rothschild (1868–1937) that we now turn. Lionel Rothschild was not robust as a child, and so he was home educated. He had a passion for nature that had developed early on while at his home, Tring Park in Hertfordshire, UK. His parents had moved there two years before he was born and were in the process of rebuilding. A joiner named Alfred Minall, whose job was to work with the wooden structures, dabbled in taxidermy, a hobby that one day young Lionel observed. Miriam Rothschild wrote in her book about her uncle *Dear Lord Rothschild: Birds, Butterflies and History*, that

Illustration by Thomas William Wood based on Wallace's description of the star orchid and its pollinating moth.

during a nursery tea once 'He stood up and made, for a seven-year-old, a long and crystal-clear pronouncement: Mama, Papa. I am going to make a museum and Mr Minall is going to help me look after it.'

And guess what, he did. For as Miriam writes further, he 'amassed the greatest collection of animals ever assembled by one man'. And what started off as a corner in a shed on Albert Street, in Tring, exploded into an enormous collection that included 144 giant tortoises, 300,000 bird skins and more importantly 2.25 million butterflies and moths… well he did start collecting and mounting them when he was eight. In 1938, following his death the year before, these along with many more specimens as well as his books and correspondence (and more!) were the single largest gift ever to have been given to the British Museum, and the building and many of his specimens now forms the Natural History Museum at Tring. Before the collection become part of the Natural History Museum, young Rothschild was a regular visitor to South Kensington, and at the age of 13 this precocious zoologist was enthralling the then Keeper of Zoology, Albert Günther (1830–1914). A fruitful friendship developed where Günther provided Lionel with an education in natural history and encouraged him with his museum. Museums, collections and Lepidoptera were to become the dominant passions in his life (with the odd scandal thrown in).

After a brief sojourn as a banker, it become clear to all that Lionel's path lay with natural history, as he grew his collection through both his own expeditions as well as funding collectors across the globe. In 1903, he and curator Heinrich Ernst 'Karl' Jordan or K J (1861–1959) published their tome *A Revision of the Lepidopterous Family Sphingidae*. So far so what, but it's here that Darwin's long-awaited long-tongued moth was first described, initially as a subspecies *Xanthopan morganii praedicta* (Rothschild and Jordan, 1903). Sadly, nothing is known about the collector or collectors, or exactly when and where it was collected. All we know is that it was from Madagascar. As a museum curator nothing is more annoying than lack of data and sadly many specimens are devoid of it, especially the Lepidoptera. This might be in part explained by the propensity of some collectors to be more interested in pretty animals than furthering scientific endeavour.

I asked the Carnegie Museum of Natural History in Pennsylvania, USA – who look after the actual specimen that was used by Rothschild and Jordan – for more information. The specimen is labelled ambiguously at best, which is a shame for many reasons, but it validated Darwin's original musings. And *X. morganii praedicta* has now been promoted from subspecies to species, thanks to a discovery by my colleague Dr David Lees in 2021, collaborating with a team led by Professor Joël Minet. 'I unrolled and measured the proboscis of a male [hawkmoth] in the Madagascan rainforest, realising that it was probably the global record

---

b. P. morgani praedicta subsp. nov.

♂♀. Breast and abdomen beneath with an obvious pinkish tint. Upperside of body and forewing, and underside of wings also somewhat pinkish. Black apical line of forewing, extending from costal to distal margin, broader than in the preceding, black discal streak R³—M¹ also heavier.

Hab. Madagascar.

Type (♂) in coll. Charles Oberthür ; a *female* specimen in coll. Mabille.

Wallace, in *Natural Selection*, p. 146 (1891), speaking of the adjustment between the length of the nectary of orchids and that of the proboscis of insects, says : " In the case of *Angraecum sesquipedale* it is necessary that the proboscis should be forced into a particular part of the flower, and this would only be done by a large moth burying its proboscis to the very base, and straining to drain the nectar from the bottom of the long tube, in which it occupies a depth of one or two inches only. . . . I have carefully measured the proboscis of a specimen of *Macrosila cluentius* from South America, in the collection of the British Museum, and find it to be nine inches and a quarter long ! One from tropical Africa (*Macrosila morgani*) is seven inches and a half. A species having a proboscis two or three inches longer could reach the nectar in the largest flowers of *Angraecum sesquipedale*, whose nectaries vary in length from ten to fourteen inches. That such a moth exists in Madagascar may be safely predicted, and naturalists who visit that island should search for it with as much confidence as astronomers searched for the planet Neptune,—and I venture to predict they will be equally successful."

As the tongue of *P. morgani praedicta* is long enough—about 225 mm. = 8 inches—to reach the honey in short and medium-sized nectaries of *Angraecum*, the moths will not abandon the flowers with especially long nectary without trying to reach the fluid, which fills up, in hot-house specimens of *Angraecum*, about one-fourth of the nectary. The result would be that flowers with exceptionally long nectaries would be as well fertilised as such with short nectaries by a moth which could reach the fluid in the long nectaries only when a greater quantity of nectar had collected. *X. morgani praedicta* can do for *Angraecum* what is necessary ; we do not believe that there exists in Madagascar a moth with a longer tongue than is found in this Sphingid.

Original description of *X. morganii praedicta* (Rothschild and Jordan, 1903) (incorrectly written as *P. morgani praedicta* in the description).

The original type specimen from Carnegie Museum of Natural History of Wallace's sphinx moth, *Xanthopan praedicta*.

holder,' says Lees. 'The taxonomic change we now propose finally gives long-deserved recognition, at the species level, to one of the most celebrated of all Malagasy [insects]'. As well as being a species in its own right, the moth has been renamed Wallace's sphinx moth, *Xanthopan praedicta*.

This hawkmoth is by no means unusual in having exceptionally long mouthparts. Aphids, those plant-sucking true bugs (Hemiptera) in the genus *Stomaphis* (Walker, 1870), have extremely long mouthparts, but within these species they have a very long labium (lower lip) and piercing stylets that are twice their body length, enabling them to penetrate the thick bark of trees! Or what about the horsefly *Philoliche longirostris* (Hardwicke, 1823) found flying around the Himalayas, whose proboscis has been recorded to be around 60 mm (2¼ in). That's impressive. But the longest insect mouthpart in relation to its body is found in another species of fly, *Moegistorhynchus longirostris* (Wiedemann, 1819), part of the Nemestrinidae. As with Wallace's sphynx moth, and many that are termed the long-tongued flies, this species lives on the African continent, but this time South Africa. Its proboscis can be up to eight times its body length with some specimens having a mouthpart 83 mm (3¼ in) long.

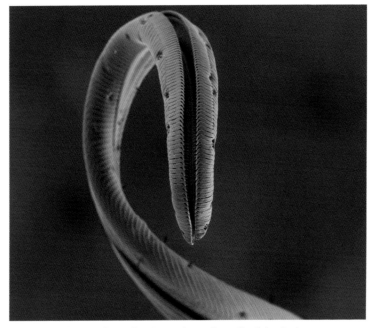

The two maxilla joined together that make up the walls of the drinking straw more formally called the proboscis.

However, the proboscis is actually far from a simple straw. Lepidoptera often drink from wet soil or rotting fruit, or suck up droplets of nectar from a surface, so they must take liquid in while limiting the entry of surrounding debris. So how do they do that? Adler and Kornev have found that the hawkmoth proboscis has tiny pores along its length, with complex sponge-like properties to draw up fluid. Not only that, the tube simultaneously carries out an essential self-clean brought about by a mosaic of attractant and repellent features that Adler describes beautifully as akin to 'valleys and ridges'. If the former are hydrophobic and the latter are hydrophilic, then you can get an overall flow of liquid into the proboscis and then up and along, without leaving any residue behind and without sucking up anything the moth doesn't want. Adler, Kornev and their team have been studying the proboscis as a blueprint for a micro-probe that uncurls, bends and takes up tiny amounts of liquid in much the same way as the moth's proboscis. Imagine a tiny probe for forensic work that is immune to contamination. Imagine being able to reverse the uncontaminated flow, creating reusable needles for mass vaccinations. And the team's ambition goes further – they want a probe to be able to take fluid out of a single human cell, which is 10 times smaller than the diameter of a human hair. The ultimate goal is to develop what they call fibre-based fluidic devices, among them probes that could eventually allow doctors to pluck a single defective gene out of a cell and replace it with a good one. The development obstacles they face are due to the incredibly small sizes they are dealing with. The research, therefore, has not been as free-flowing as the subject itself.

There are so many possibilities to this work, yet mimicking what has taken more than six million years of evolution to produce, isn't going to come quickly. Amazingly it wasn't until 1992, 130 years after Darwin's initial prediction, that scientists finally observed and filmed Wallace's sphinx moth visiting the star orchid. A male hawkmoth carrying its little pollen packet was recorded, and photographs using night vision equipment were taken – it wasn't direct evidence of feeding, but it was close. Then in 2004 a video of the act of pollination was taken in all its extraordinary detail. Being able to see this must have been pure joy to the scientists who recorded it, witnessing the 25 cm (10 in) tongue uncurling on the approach to the flower's narrow nectary, to see it slow down, the dextrous mouthpart inserted whilst hovering at the flower's opening, and then to drink with amazing precision.

The star orchid and the sphinx moth relationship was typical of the kind of research that the theory of natural selection highlighted – Darwin hadn't just been collecting, classifying and describing but had turned natural history from a science of observing into a science of predicting. When writing the second edition of his orchid book, he wrote about entomologists mocking him for suggesting that a moth with such a long tongue existed. But he'd stuck to his guns, convinced that nothing else could explain what everyone would observe, and was eventually proven right – a lesson for us all.

CHAPTER THREE

Drosophila melanogastronauts

Am not I
A Fly like thee?
Or art not thou
A man like me?

The Fly, William Blake (1794)

E NGLISH POET William Blake had hit upon the truth in his piece *The Fly*, written more than 200 years ago. We are indeed very much like flies, but we had to wait a further hundred years to realize it and start taking advantage of this fact. You may not think you have much in common with them – after all, they have wings, six legs and weird eyes to name but a few physical differences. But as our mothers told us, it is what is on the inside that counts. It's all about the genetics, for we share many similar genes – in fact 75% of human disease-causing genes are shared with the humble vinegar fly *Drosophila melanogaster* (Meigen, 1830). It is one of the most important and most studied species on the planet. Despite this, left to their own devices these flies would be content to spend their adult lives doing nothing much more than hovering around a pile of rotting apples, getting drunk and indulging

The best mix-tape there is. A cassette of multiple *D. melanogaster* prepared to be sent into space.

Super model extraordinaire – *Drosophila melanogaster.*

in one-night stands. But fate intervened for both us and the vinegar fly back at the start of the twentieth century.

Much has been written about this species and our working relationship with it. Stephanie Mohr is a genetist at Harvard Medical School, and has spent most of her career working with these flies, publishing *First in Fly: Drosophila Research and Biological Discovery* in 2018. In her lab, which is like thousands of fly labs across the globe, she has trays and trays of these flies that are set up for genetic studies. She readily admits that, although they do not look anything like humans, their overall body plan is the same, and they share all the senses that we have – they touch, they taste and so on. And internally we see this too. They have a heart, and it beats. They have a gut that processes and excretes food just like ours. And they have an astoundingly sophisticated brain given their size (3 mm or 0.12 in long). And this, she writes, has enabled us to study many human processes at a simpler level than if we were to study them in ourselves.

The vinegar fly decided early on in its evolutionary pathway that it very much liked hanging around us, or rather our food – what we term an anthrophilic species. It's widely thought to have originated from equatorial Africa where wild populations have been found to be closely associated with the abundant South African marula fruit, *Sclerocarya birrea* (A. Rich) Hochst. Recent research by Marcus Stensmyr and collegues in Zimbabwe found that humans in prehistoric times brought this fruit deep into their caves – as evidenced by images captured in exquisite cave paintings and the discovery, in dark cave recessss, of the remains of husks. It's likely that these fruits would have ripened then fermented as they began to rot, and the smell of fermentation would have attracted *Drosophila* inside. Stensmyr and colleagues carried out a series of trapping experiments, and in recreating these conditions found that of all the species of *Drosophila* it was *D. melanogaster* that was drawn into these fruit-laden dark caverns. So a series of choices and chance events tens of thousands of years ago led to us having this minute companion that began following us around as we moved across the planet – hitch-hiking a lift on slave ships out of Africa, and from the Caribbean, along the trade routes of sugar and rum (a lovely tipple for the flies), and as stowaways amidst the new traffic in bananas and fresh fruit that emerged after the American Civil War.

The experimental design. The larva of *D. melanogaster* in a genetics lab.

We leave food around – enticing vinegar flies to come closer. That bowl of fruit may look pretty to you but it acts as an all-in-one 'dinner and day care' for their maggots. Amusingly this species of fly arrived in North America just a few years before researchers would begin to adopt this organism as a true biological footsoldier, when in 1875 the New York State entomologist Joseph Albert Lintner reported they had been 'bred from a jar of pickled plums'. Twenty-five years later it was described as one of the most common species in the USA. *Drosophila melanogaster* is a true fly (Diptera) in the family Drosophilidae, the vinegar flies. Many folks will know it as the fruit fly, but true fruit flies are in the family Tephritidae. Tephritidae contains some of the serious economic pests but the majority of the Drosophilids are well behaved. They aren't biological control agents, medically important or anything else that may assist or destroy us. So why did we pick them as a model organism to study?

Flies were not the first model organism. Previous examples include guinea pigs, hares and sheep. Especially with the last on this list, these were not small animals to house in labs and their generation times were long. The guinea pig, the smallest of these models, births on average two to four pups at a time, with the gestation period lasting around two months. They live on average four to eight years so let us say six for this story, which means a female can (allowing for the four-week development period, being permanently pregnant and reproducing till death) produce 72 to 144 pups. That is a lot in comparison to humans, but it is nothing on this fly – they can produce an equivalent number in a couple of weeks. And it's not just their reproductive prowess that makes them great model organisms but also the fact they are just 3 mm in length! You don't need a lot of space or money to rear these animals.

The dramatic leap of faith to use these flies as the laboratory darling began at the start of the twentieth century by the Harvard University scholar Professor Charles Woodworth (1865–1940). It is not clear how he came to breed them, but the ease with which he did, and their short generation time must have been appealing qualities. Woodworth then recommended them to his colleague William Castle, who initially worked on mammals but utilized the flies to study inbreeding. During this same period another entomologist, Dr Frank Lutz (1879–1943) at

Thomas Hunt Morgan's famous Fly Room at Columbia University. Note the bananas that were used to feed them.

the American Museum of Natural History, also began studying this fly's basic biology, publishing more than a dozen papers about it. It was from Lutz that Professor Thomas Hunt Morgan (1886–1945) introduced them to his own lab at Columbia University (that he had joined four years previously in 1904). And it was in this lab that these tiny creatures (and Morgan to be fair) became megastars.

Morgan was not an entomologist, he was an embryologist. His interest was not in the flies *per se*, but in development and heredity. He was always keen to find new organisms to work on, anything from pigeons to rabbits to frogs, snails and a host of other creatures. His wide tastes were often a source of amusement amongst his colleagues, who joked 'he has more irons in the fire than an ordinary man has coals'. And so he squeezed flies into his small and already overcrowded lab. As science historian Jim Endersby has stated, flies were perfect for an academic biology lab since they're plentiful at the start of the academic year due to a ready supply of autumn fruit, and breed easily in a warm lab throughout winter, producing a new generation every couple of weeks.

By the time *Drosophila* crossed the threshold from field to Morgan's dingy lab on Manhattan's Upper West Side, biologists had begun to appreciate for the first time the long-neglected genetic experiments of the nineteenth-century Austrian Gregor Johann Mendel (1822–1884). Mendel was a monk, a mathematician, a meteorologist, a physicist and a botanist. And these diverse fields combined to enable him, thanks to the help of peas, to establish the laws of classical genetics in 1865 – that is the seeing of changes in the offspring from that of the parents brought

Mendel's scribbled notes from his ground-breaking pea experiments.

about by reproduction. I have my father's eyes and my mother's nose (not literally thankfully) but in many ways I am physically different from them (many ways…). And it was Mendel who identified these basic mechanisms of heredity by playing around with common garden peas *Pisum sativum* (Linnaeus, 1753), crossing them again and again to see which pea traits, for example colours or wrinkling, were passed on to the next generation. Mendel's ground-breaking and meticulous experiments were largely ignored at the time of publication. Darwin was sent a copy but never read it, a fact that seems quite amazing considering his work on the theory of evolution by natural selection was looking at the changes in heritable traits, be they physical, as studied by Mendel, or behavioural.

On the Origin of Species by Charles Darwin had been published a few years before Mendel's work in 1859, and the theory of evolution developed by Darwin and naturalist Alfred Russel Wallace was still being hotly debated. As a scientist who valued experimentation over observation, Morgan was sceptical of Darwin and his theories, and he was hugely sceptical of Mendel's laws of inheritance – indeed he stated, 'In the modern interpretation of Mendelism, facts are being transformed into factors at a rapid rate. If one factor will not explain the facts, then two are invoked; if two prove insufficient, three will sometimes work out.' Morgan began breeding the fly to test an alternative to the theory of natural selection, which he felt was not sufficient to explain the origin of new species. He started breeding *Drosophila*, not for Mendelian studies, but as a foray into laboratory-based experimental evolution. He'd been drawn to the Dutch botanist Hugo de Vries' new theory that species evolved through discrete jumps, which de Vries called mutations. De Vries (1848–1935) had predicted that, under certain conditions, animals can enter 'mutating periods'. With *Drosophila*, Morgan wanted to see if he could induce such a mutating period through intensive inbreeding, in effect re-creating evolution in a test tube. And as the turnover of *Drosophila* in his tiny lab grew and grew, it gradually transformed into the Fly Room.

Science historian and geneticist Professor Matthew Cobb, University of Manchester, UK, has been using *Drosophila* for many years as lab organisms, but focused on their maggots to understand their physiology, for instance how do maggots smell (please don't say awful), and has published many a tome on the value of these flies to the ongoing scientific endeavour. Cobb explains that Morgan was looking for those examples of sudden change that he believed underpinned de Vries' mutation theory in the way new species evolved, and this was most easily done with flies.

For two years Morgan persisted in breeding these flies, but as he wrote he 'got nothing out of it'. His frustration was mounting, especially as this was an astonishing period of scientific endeavours and insights published by his peers. Chromosomes had been observed in the late 1800s – the parts of the nucleus that contain the deoxyribonucleic acid or more simply DNA that form the genes – but we didn't know that then.

D. melanogaster doing what they do best.

What we did know about was inherited traits, thanks to the rediscovery of Mendel's work, but again this was just concept – we did not know how or where in the cell it happened. In 1902, German zoologist Theodor Boveri (1862–1915) was working on embryonic development and saw that in two species of sea urchin, *Psammechinus microtuberculatus* (Blainville, 1825) and *Sphaerechinus granularis* (Lamarck, 1816), chromosomes were needed for this development to happen properly. That same year across the ocean and back to Colombia University, American geneticist Dr Walter Stanborough Sutton (1877–1916) observed the chromosomes splitting during meiosis – cell division during reproduction – to form daughter cells. Together these two findings led to the Boveri–Sutton chromosome theory, which identified these once mysterious shapes in the nucleus as the genetic material responsible for Mendelian inheritance. The behaviour of this genetic material appeared to mirror the evidence from Mendel's breeding experiments. Mendel's work was on its path to vindication.

Morgan's lab, meanwhile, had been a hub of activity trying to prove/ disprove theories and it now housed many bottles of flies that he and his team were working on, trying to get some answers to these evolutionary questions. But after two years of failed attempts to prove the de Vries mutation theory, by subjecting flies to various diets and chemicals to induce a 'mutation period', he was on the brink of losing interest. That was until one day in 1910, when amongst the stock of normally red-eyed flies, a white-eyed male was born. A mutant. Something had happened – a de Vriesian mutation period perhaps? To find out what had caused this sudden change in colour, Morgan set up a simple breeding experiment.

At first Morgan was confused, since this mutant change wasn't as large as de Vries' theory had predicted. In addition, the first generation mutant offspring could all be crossed with each other so didn't constitute a new species; when a normal red-eyed female was crossed with one of these white-eyed male mutants, all the offspring appeared to have red eyes. But when this second generation was crossed, red-eyed and white-eyed flies appeared in a ratio of three to one. Let's call red R and white w. The mixing of the male and female chromosomes would involve a combination of RR, Rw, and ww. Red eyes would appear dominant in this situation and so crosses that resulted in RR and Rw would always produce red eyes, but parents who both had the recessive ww would produce white eyes. So, the mutation, far from resulting in a new species, was in essence evidence of a classic Mendelian recessive factor.

Most important of all, none of the females had white eyes. Morgan initially proposed that the gene may be lethal in females and was killing them early. But thanks to the work of his brilliant PhD student Nettie Stevens (1861–1912), and with help from the previous head of the department Professor Edmund Wilson (1856–1939), sex chromosomes were determined – the X and the Y chromosomes. These are the sex cells that each contain half the genetic material of their parent, so Morgan set about trying, successfully as it turned out, to determine that the white-eye trait was indeed an inherited sex-linked character. This seminal finding led Morgan to conclude that some genetic traits are not inherited independently (as Mendel had supposed) but rather must be linked. This was ground-breaking and slightly controversial (it contradicted one of Mendel's rules) but as the Harvard geneticist Mohr points out there was data to back up his conclusions; more than

A red-eyed female and white-eyed male *D. melanogaster*.

4,000 flies were studied in determining this. This level of replication could not have happened with other model organisms, and so not only were the results revolutionary in our understanding of genetics but also transformative in establishing this fly as the model organism. With these results Morgan not only realized he'd been wrong about the de Vries mutation theory, he also confirmed the chromosome theory of inheritance and linked for the first time chromosomes, genes, mutations and inheritance.

Morgan and his team went on to notice more and more variants – for instance different shaped bristles or notches in their wings. They went on to find out where on the chromosome these were located, and came to the conclusion that those characters more often inherited together were found physically closer on the chromosome. In 1915, Morgan and his lab team published *The Mechanism of Mendelian Inheritance* summarizing all these discoveries with evidence backed up by huge sets of data. They may not have been the first to suggest that hereditary or Mendelian factors were located on chromosomes, but their claims were compelling – particularly since Morgan happily donated stocks of the relevant flies for anyone who wanted to check the results for themselves.

Modern genetics had been given life with the birth of that first male fly and Morgan would go onto receive a Nobel Prize in 1933, one of six that in recent decades have been awarded to researchers working on *D. melanogaster*. These studies at the Columbia Fly Room ushered in a new era of research, launching experimental genetics to the world, and in doing so, propelled the humble vinegar fly to one of the most influential models on the planet.

That first gene mutation was named 'White' and for most part the naming of genes has a kind of backward logic in which genes get named after the mutant version of the gene – most of the eyes were normally red. We know now that there are 14,000 genes on four chromosomes in this *Drosophila* species (humans in comparison have between 20,000 and 25,000 genes and 23 paired chromosomes). That is a lot of names, and possibly to relieve the lab workers' drudgery of the often-repetitive nature of their work, scientists have dreamt up some truly silly names. There is the Ken and Barbie gene, whose mutation results in flies without external genitalia. The tin man gene is as sad as it sounds as they are born with no heart, while the Swiss cheese gene results in cavities in the brain, resembling this holey fromage. Flies carrying the mutant (variant) of the Groucho gene have extra bristles above their eyes resembling the bushy eyebrows sported by the American comedian Groucho Marx. There is the lush gene, and the cheap date gene, the former often causing flies to be attracted to ethanol (and related substances) whilst the latter makes them very susceptible to it! And these two are interesting as they can help us understand human behaviour and alcohol dependency. Researchers back in 1998 identified the mutant gene with this increased sensitivity to ethanol using an inebriometer – I promise I did not make that name up. This terminology may seem a bit silly but the research isn't, as alcohol is not only one of the most widely used drugs in the world but there are many medical as well as social issues associated with its misuse. Just in England there are nearly 600,000 dependent drinkers (2017/2018). So, in trying to determine the roles that genes play in addiction, we can develop ways of regulating the impact of these genes. This is moving away from just physical traits being manipulated and altered, to the study of behavioural traits as well.

Drosophila have for decades been perceived as being in lab vogue. But that hasn't always been the case. There's an evocative image of a giant wooden model of a *Drosophila* being proudly held by a scientist at the California Institute of Technology (Caltech) in the early 1970s. It was perhaps a homage to the fly's past glory days – since by this time, bacteria and viruses, being yet smaller and simpler, had begun to replace the flies as the model organism, with the bacterium *Haemophilus influenzae* (Lehmann and Neumann, 1896) winning the award for first whole genome to be sequenced back in 1995. We would have to wait another five years for the *Drosophila* genome to be sequenced. *Drosophila*, although still common in labs was losing out in the experimental department to these simpler, and even cheaper, rivals.

But the American Professor Seymour Benzer (1921–2007) (pictured gazing at the wooden model) had a great deal of faith in this species. Not only could it perform many sophisticated behaviours that we humans do – such as learning, courting and keeping time – but it could walk on ceilings and fly. He wanted to know more about how they were able to do this rather than just exploring the genes that controlled their physiology or morphology. And so Benzer embarked on pioneering work on genes and behaviour that reignited our passion for this creature. He was given a microscope for his thirteenth birthday that 'opened up the whole world', an experience I can relate to, having been given one at a similar age. He was to develop an interest in this microscopic world to a much greater benefit than me who would just get excited at watching fleas jump (see chapter one). Benzer initially trained as a physicist, but he moved onto genetics, inspired by another physicist Edwin Schrödinger (1887–1961), and was the first to determine that mutations can occur at different sites on the same gene. In doing so, he showed that genes were not comprised of indivisible units, as had been previously thought, but by a linear row of chemical blocks called bases – imagine a row of multicoloured Lego pieces stacked along the ground and you have the

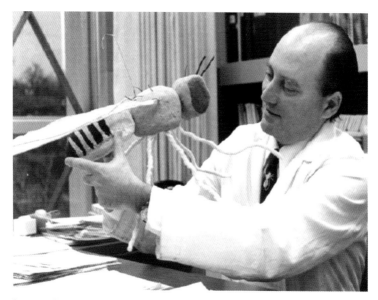

Seymour Benzer with a *Drosophila* model in his office at the California Institute of Technology in 1974.

right idea. And even a change in a single brick or more technically put, a single base pair, could result in a mutation. Benzer's understanding of these bricks formed the foundations of molecular biology.

Not satisfied with stopping there, however, he turned his restless spirit to look at *Drosophila* while on a sabbatical at Caltech, and again broke new ground, this time in founding the modern field of neurogenetics – how genes influence behaviour. Previous attempts to investigate this phenomenon had included studying how complex behavioural traits could be altered by intercrossing or enhanced by selective breeding. But Benzer had a different approach in mind – he would look for a model organism in which it would be possible to define single-gene mutations that altered specific behaviours. When he chose *Drosophila* as his laboratory tool many of his colleagues were unconvinced. But Benzer was undeterred. He once said, 'If everyone you talk to says you shouldn't do something, you probably shouldn't do it, and if everyone says you should do something, you should also probably not do it. But if half the people you talk to tell you to do it and half say you're crazy, then you should definitely go ahead.'

Benzer turned his sabbatical at Caltech into an indefinite stay, and in determining how genes were linked to behaviours he was able to show that these little animals were much more sophisticated than we had initially given them credit for – including an ability to keep time. In 1973 Benzer published a paper in which he described a circadian rhythm mutation in the fly – they had an internal clock or circadian rhythm just like we do, but a mutation Benzer had created almost certainly meant there was a single base of DNA that had been altered. It produced either a fly with no clock at all that never went to sleep at any reliable time, or a fly that woke up very early and had a clock that was running at about 17 hours rather than 24, or a fly with a clock that ran at about 27 hours. Benzer named this mutation 'period'. And it is the same system in us as in a fly. Whilst it is more complex in mammals, it is essentially the same gene and it works in basically the same way. So, Benzer had used these flies to understand circadian rhythms, and he would go on to investigate aging and its reversal, and even a fly that couldn't learn – that last one feels very familiar.

From the early days of just identifying genes, the value of these flies in many different disciplines had now expanded to enable us to study areas such as neurodegeneration, cancer and sleep to name but a few subjects. And this led to studies not just on Earth, but above it, too – this species was the first animal to be sent into space. On 20 February 1947 on board a V2 rocket, flies were launched from a missile range in New Mexico, USA, and became the very first astronauts. These explorers travelled 108 km (67 miles) into the air before parachuting (the rocket not the individual flies) back down to Earth. In doing so, they travelled 1.5 km(1 mile) into what the American space agency NASA recognizes as the starting point of space. The idea of sending a flying animal up in a rocket still amuses me more than it should do.

Flies have been on many a space mission since. There is even a Fruit Fly Lab run by NASA. Since 2014 the Fruit Fly Lab on the International

Tiny little astronauts in their tiny little spaceship!

Space Station has been the setting for a series of experiments looking at the effect of long duration space missions, on everything from body clock gene expression to body defence systems. Dr Sharmila Bhattacharya, a neurobiologist, leads a lab at NASA's Ames Research Centre and since 2015 has been involved in sending flies on the shuttle into space. They are not in little suits sadly but in tiny cassettes the size of a deck of cards, with many thousands being sent up confined to a space smaller than a bread bin. Once more it is their ability to breed rapidly, together with their minimal housing requirements, that make them ideal for research into space travel, to determine the impact such journeys would have on us larger travellers.

Bhattacharya and many other researchers have been studying how flies adapt to living in space, and their value lies in the fact that you can undertake multigeneration studies in a short span of time. This allows us to understand how flies at all stages of their lives adapt to extra-terrestrial living and undergo physiological changes such as alterations in sleep habits under these alien conditions. Sleep, as many of us will testify, is a very important process. It is while we sleep that our body and brain undergo a period of cleaning, repairing and resting, essential for maintaining their proper function – we have all experienced a foggy brain in the morning when our sleep has been disturbed. And it is not just the impact of poor memory and grogginess, but also the effect on our immune system. This combined with weightlessness, properly called microgravity, has some clear implications on prolonged space flight.

The first insights into the impact of space on heart function is the latest study on 'astroflies', published by Bhattacharya and colleagues in 2020. These flies were born in space and spent the first three weeks of life weightless, thanks to microgravity – the human equivalent of three decades. Upon arriving on Earth for the first time, these 'aliens' were quickly transferred into appropriate laboratory settings to ensure that the effect of gravity had no impacts on findings. Researchers measured their heart rate and contractile ability by making them climb up test tubes – a tiny fitness test – and by comparing these with the regular Earth flies they discovered that astroflies had smaller hearts that pumped less. Furthermore, there was a change in the arrangement of and a reduction in the muscles that caused the contractions. This is not only important for understanding what happens to our bodies in space but also in our everyday lives, as this is the exact opposite of what happens to our hearts after we've had a heart attack. The next step is to find out what proteins are affecting the muscle development.

We have come a long way since Morgan brought *Drosophila* from our fruit bowls into the lab. In doing so he turned the fly from a living organism into what is effectively a piece of lab equipment. He utilized *Drosophila's* small size, its high fecundity and short life cycle, the low cost and ease with which it can be cultured, its small chromosome complement, and its ability to withstand mutation and crossbreeding experiments. A century on, for any would-be or card-carrying drosophilists, the tools of the trade remain the same. As one guidebook for students jokes, 'the fly frequently requires you to do an apprenticeship on any important project. They will not start to perform until they are certain you are serious'. Again and again the fly has proved itself to be a powerful system for study, exposing new information on human genetic disease and fallibility simply by doing what it does best – eating, drinking and mating… an enviable life.

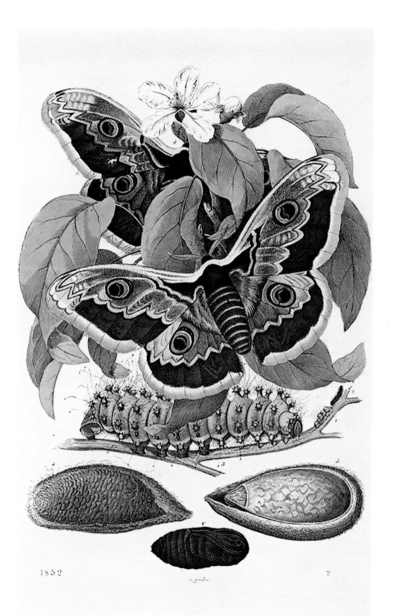

CHAPTER FOUR

Cycles of change

The Silk-worm digs her Grave as she doth spin,
And makes her Winding-sheet to lap her in:
And from her Bowels takes a heap of Silk,
Which on her Body as a Tomb is built:
Out of her ashes do her young ones rise:
Having bequeath'd her Life to them, she dyes.

Lady Margaret Lucas Cavendish – 1671 (2nd edition)

I̲T̲ ̲I̲S̲ ̲H̲A̲R̲D̲ to believe that just over 350 years ago, it was widely thought that insects spawned spontaneously from dust or mud. In 1671, the rather influential and illustrious Lady Margaret Lucas Cavendish, Duchess of Newcastle-upon-Tyne (1623–1673), described as a philosopher, a writer, a scientist and a poet, published a verse of the dramatic 'process' – a butterfly rising out of the decay of a dead caterpillar. Nowadays we know this is not the case, but that does not mean the truth is any less dramatic. Their extraordinary life cycles, their ability to metamorphose from grub to flying adult, has enabled insects to conquer nearly every terrestrial habitat.

Insects are found in many freshwater ecosystems, have ventured into the marine environment, and let's not forget about the *Drosophila* and their space travel. Twenty per cent of all the insects develop by simply

Wood engraving, dated 1852, of the great peacock moth, *Saturnia pyri* (Denis & Schiffermuller, 1775).

undergoing a series of moults until they reach the adult form, but around 280 and 300 million years a radical change in development evolved, and today the huge majority go through what is arguably the most amazing thing to occur on this planet – what appears to be an almost complete change in body form. These insects undergo a complete metamorphosis, a process that takes its name from the Greek for transformation. Just in numbers alone, 65% of all animals species on the planet undergo metamorphosis.

The enormous numbers of metamorphosing insects on the planet speak for its success as a reproductive strategy. And while it has long inspired misunderstanding and mysticism, and remains in part a biological mystery even today, it is taking on a new environmental significance. That may seem a bit sensationalist but then isn't it pretty sensational that a bright green caterpillar can change into a hirsute flying emperor moth, or that a writhing white maggot can transform into a metallic flying bluebottle? Ask around and few people have given thought to what actually happens during this period of change – this hidden part of an insect's life. To understand it we must first understand the different stages of life as an insect.

The four main orders of insects – the Lepidoptera (butterflies and moths), Coleoptera (beetles), Diptera (true flies) and the Hymenoptera (bees, wasps, ants and sawflies) are all holometabolic insects, which means they undergo complete metamorphosis, where the majority go from egg, to larva, then pupa to adult. Most folks only think about adults as insects, disregarding the three, often longer, stages that need to be completed before the grand, adult finale. Even if you have thought about the larvae of these species, and as a gardener you definitely would have, have you thought about what exactly is going on in there? This bizarre life cycle has been pondered by many for a while now.

Ideas of how insect life emerges had prevailed for centuries, with few linking the larval form with the adult. The Ancient Egyptians believed that honey bees, *Apis mellifera* (Linneaus, 1758) were the transformed tears of the sun god Ra that had fallen to Earth. They were the first bee keepers, maintaining the bees, rather than just removing honey and other products, during the reign of King Senusret III (1870–1830 BC).

a = He of the sedge and the bee

b = sealer of the King of Lower Egypt

c = bee

d = honey

e = beekeeper

f = beekeeper

g Chief beekeeper of Amun

Bee-related hieroglyphs.

There is evidence of the Egyptians using smoke, albeit accidentally at first, to calm bees to remove the golden honey. All of this was detailed in their hieroglyphic accounts, but what was not detailed was their thoughts on the honey bee's life cycle.

The Greco-Romans revered the honey bee as much as the Egyptians had done, and much was written about them and their insect relatives. Aristotle (384–322 BC), so famous he only needs one name, was a believer in spontaneous generation, the scientific theory that living creatures actually arose from non-living material, and was one of the earliest scholars to articulate this theory. Yes, fleas from dust, or maggots from dead flesh, as long as the material contained 'pneuma' what he described as 'vital heat'. He was not alone. In the *Bible*, God tells Moses to 'stretch out your staff and strike the dust of the earth, it may become gnats through all the land of Egypt' – a great feat indeed.

ATHANASII KIRCHERI
E Soc. Jesu

MUNDUS
SUBTERRANEUS,
In XII Libros digestus;

QVO

Divinum Subterrestris Mundi Opificium, mira
Ergasteriorum Naturæ in eo distributio, verbo πανταμορφον
Protei Regnum,

Universæ denique Naturæ Majestas & divitiæ summa
rerum varietate exponuntur. Abditorum effectuum causæ acri indagine
inquisitæ demonstrantur ; cognitæ per Artis & Naturæ conjugium ad
humanæ vitæ necessarium usum vario experimentorum apparatu,
necnon novo modo, & ratione applicantur.

TOMUS I.

AD

ALEXANDRUM VII.
PONT. OPT. MAX.

AMSTELODAMI,
Apud JOANNEM JANSSONIUM & ELIZEUM WEYERSTRATEN,
ANNO cIɔ Iɔc LXV. *Cum Privilegiis.*

Mundus Subterraneus, a how-to for making insects, frogs, snakes
and scorpions at home.

This theory of insects being generated from non-living material persisted for a long time. Athanasius Kircher (1602–1680), was a German Jesuit scholar, and one of the first to adopt the new-fangled instrument called the microscope and use it to study some of the very small stuff, including microbes. This didn't stop him from creating a recipe book for all animals that were not thought to breed – the lower beings that included frogs, snakes, scorpions, and insects. First published in 1664, *Mundus Subterraneus*, was to be one of Kircher's most popular publications.

Want to know how to make flies from scratch? Well, here is Kircher's recipe (translated by Gottdenker 1979: 576, from the 1668 edition), 'Collect a number of fly cadavers and crush them slightly. Put them on a brass plate and sprinkle the macerate with honey-water. Then expose the plate, as chemists do, to the low heat of ashes or of sand over coals, or even of horse dung; and you will see, under the magnifying power of the microscope, otherwise invisible worms, which then become winged, perceptible little flies and increase in size to animated full-fledged specimens.'

This amusing theory of spontaneous generation, held by him and many others at the time, was finally overturned thanks to the Italian physician Francesco Redi (1626–1697). Although more a parasite man than an insect one – he is described as the father of modern parasitology – he was important in debunking many of the inaccurate scientific beliefs of his era, including spontaneous generation. In 1668, Redi published *Esperienze Intorno alla Generazione Degl'insetti* (Experiments on the Generation of Insects) in which he both predicted and proved that adult flies were needed for maggots to 'appear' or more correctly to be found on a food source (in this case dead snakes), 'the putrescence of a dead body, or the filth of any sort of decayed matter engenders worms'. Redi was unusual in that he believed in the experimental method and stated that 'in order to verify observations, we frequently approach or recede from the object that we wish to examine, change its position or its light'. He was thorough in his approach and questioning of what had been previously proposed and accepted.

Reading Redi's description of his investigation into how flies lead to maggots is akin to being there, as he writes with such curiosity. You want to find out what happens next! First the dead snakes were placed in an open box, and it wasn't long before he witnessed these 'worms' (aka the maggots) feeding, growing and then, interestingly, escaping. Intrigued, he started again but this time once he witnessed the maggots as he sealed the box so they could not escape. He then described how these encased maggots ceased all movement as if they were 'asleep', and changed shape into an 'egg'. These egg-shapes darkened, eventually turning black – he does not use the term pupa. Redi transferred his black egg balls into separate glass vessels and then, on the eighth day, from each one 'came forth a fly of grey colour, torpid and dull, misshapen as if half finished, with closed wings; but after a few minutes they commenced to unfold and to expand in exact proportion to the tiny body, which also in the meantime had acquired symmetry in all its parts. Then the whole creature, as if made anew, having lost its grey colour, took on a most brilliant and vivid green; and the whole body had expanded and grown so that it seemed incredible that it could ever have been contained in the small shell.' I have sat in my garden and witnessed this event, and I recommend all of you do, for it is one of the most awe inspiring things to behold. Redi went on adding to this experiment, using all manner of animals, and more importantly comparing when he sealed the box with a cover or with gauze. The covered boxes generated no maggots, but further work with gauze-covered boxes generated both maggots and the adult flies. For adults you needed maggots and for maggots you needed adults. The deathblow for spontaneous generation had been struck, but sadly it was to be a lingering death. Redi was also known as a very good poet, and I think his sentiments were expressed quite succinctly in this little ditty about Aristotle, 'Because He's Aristotle, it implies That he must be believed e'en though he lies'.

A contemporary of Redi, Jan Swammerdam (1637–1680), a Dutch biologist and microscopist, was highly critical of all those who blindly followed the views of the Ancients and also undertook work to disprove Aristotle. Swammerdam had trained in medicine at the then New University of Leiden, Netherlands, and made major discoveries due to his careful dissections, including research into the human uterus.

But his interests and talents lay in dissecting much smaller creatures than humans. His father Jan Jacobszoon (1606–1678), had in his apothecary just south of the main docks in Amsterdam, several 'cabinets of curiosities' displaying the extraordinary and strange *naturalia* that sailors gathered from their travels far and wide. These cabinets sparked the young boy's imagination much to the concern of Jacobszoon the elder, who favoured a more stable and traditional life for his son, presumably because his own was concerned with much financial instability. Leiden University by then was the European hotspot for natural history, so the young Swammerdam, in spite of his father's wishes, began making painstaking dissections and drawings of insects – with help from some new technology, the tiny single-lens microscope. What he discovered helped to upturn the belief in Aristotle's theory of spontaneous generation, but not before someone else published work on the silkworm.

In 1669, Marcello Malpighi (1628–1694,) an Italian biologist and physicist, had also started playing with the single-lens microscope. He was to become, amongst other things, the founder of microscopic anatomy, and produced a detailed study of the silkworm titled *De Bombyce*. It was a hugely influential publication, and may not have happened if it weren't for Henry Oldenburg, Secretary of the Royal Society of London, writing to ask him, in 1667, whether he would like to communicate any of his discoveries to the society's learned members. As chance would have it, he had just finished *De Bombyce*, and so duly sent it in. Upon seeing it, the society underwrote the publication and were lavish with their expenditure – it's a large book with fold out drawings. It came four years after England's own Robert Hooke's book on microscopy, *Micrographia or Some Physiological Descriptions of Minute Bodies Made by Magnifying Flasses with Observations and Inquiries Thereupon*. Both were incredibly detailed, with stunningly accurate drawings that have withstood the test of time, but Hooke did not carry out the dissections that Malpighi did.

Swammerdam was gifted a copy of *De Bombyce*, as he was already recognized as an authority on anatomy and physiology. The exquisite dissection of many insects including the silkworm, enabled

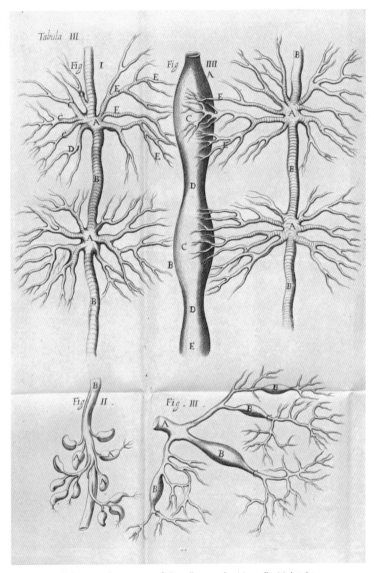

Engraving of the vascular system of the silkworm by Marcello Malpighi, in *De Bombyce*, 1669.

Swammerdam to understand insect development and transformation in a way no one had before. In his doctorate *Historia Insectorum Generalis*, published in 1669, he'd begun to create a classification system based on how an animal developed – a momentous undertaking and just under 100 years before Carl Linnaeus achieved his still widely used classification system. The first grouping, which Swammerdam termed Order, included a hodgepodge of arthropods including spiders, scorpions and ametabolous insects, those insects that showed very little change from the hatched egg stage. The second group led on from this, where the insects gradually developed from a nymphal stage to an adult. A third and fourth group adopted complete metamorphosis. Astonishingly, he was also able to show, in detail, that insects contained complex internal organs. In doing so, he too broke with the doctrine of the Aristotelian tradition, revealing metamorphosis (rather than spontaneous generation), and complexity.

His elegant observations of insect internal anatomy and of their behaviour was a robust repost to the work of his contemporaries such as Jan Goedaert (1617–1668), the influential Dutch painter and naturalist who had characterized the growth and metamorphosis of insects in *Metamorphosis Naturalis*. Goedaert proposed that butterflies came from the decay of a caterpillar and yet 'out of one and the same species of caterpillar, a butterfly is produced and 82 flyes'. This clearly added much confusion to the debate. According to Swammerdam, Goedaert was a victim of his own imagination, since he had portrayed several animals incompletely or incorrectly and his description of some insects read more like a novel than a 'true history'. Swammerdam claimed, 'The matter should be attacked not by thought but by experimentation'.

The drawings, observations and understanding of insect anatomy that came out of this period is phenomenal. The issue they all faced was that insects are small, with many organelles too tiny to see with the human eye. Simple lenses had been used to help with this but the earliest examples of having both a lens and eyepiece, what is referred to as a compound microscope, did not appear till around 1600. Who the exact inventor was is unclear – Dutch spectacle maker Zacharias Janssen is credited with one of the earliest. But one man was to become incredibly influential both for his own work and his invention of an incredibly powerful microscope (well for the time that is). Hooke describes how he refined the design of a microscope that uses three lenses and a stage light. But it was Antonie van Leeuwenhoek (1632–1723), a Dutch microbiologist, who had also been making his own microscopes and was the first to draw bacteria and protozoa (which significantly he referred to as tiny animals), that would become the more influential. Leeuwenhoek was a master builder of microscopes and opted for a single lens – of the type favoured by Swammerdam, an instrument able to magnify an object up to 300 times its size, thanks to the high-quality grinding of the glass creating a smooth and accurate lens.

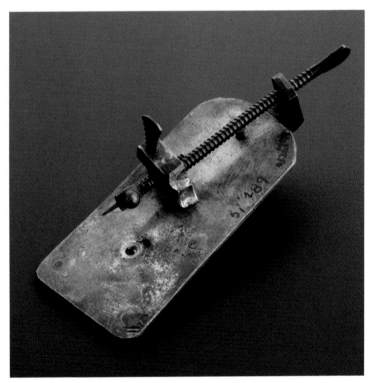

A replica Leeuwenhoek microscope.

Thanks to the new powers of this single-lens microscope, Swammerdam was able to see that the caterpillar in its pupal stage had several internal adult features before the adult emerged – the antennae, the wings, the head and even parts of the abdomen. This is an important moment for science, although errors did occur thanks to distortions caused by low quality lenses. However, it was no longer evident that butterflies emerged from the remains or mush of the dead caterpillar, but rather that the life was continuous. The animal merely changed form. Swammerdam wrote, 'The pupa not only contains all the parts of the future animal, but is indeed that animal itself ... it is nothing more than a change of the Caterpillar or worm, containing the embryo of the winged animal that is to proceed from it'.

Swammerdam demonstrated this as a party trick for the great and good in Europe, peeling the skin off silkworms – the larval stage of the domesticated silk moth *Bombyx mori* (Linnaeus, 1758) – to reveal the rudimentary wings within. He went round the world showing people how you could prove that these two apparently completely separate organisms were in fact the same one. His audiences included Cosimo III de Medici (1642–1723), the Grand Duke of Tuscany, who was touring northern Europe, in part to escape a terrible marriage but also to learn more about the scientific and artistic achievements that had been coming out of this region of Europe at the time. Swammerdam concluded, 'All the limbs of the butterfly, the fly, and other such insects, do actually grow in the worm, and in the same manner as the limbs of other animals. (…) Those parts are by no means generated suddenly and all at once, as has been supposed, but grow leisurely one after another under the skin that covers them.'

As a child of the seventeenth-century scientific revolution, these discoveries in nature reinforced Swammerdam's vague and mysterious religious beliefs rather than undermined them, to such a point that his scientific research and his religious attitudes became completely intertwined. He was convinced that the world had to be perfect because God had made it. So he was looking for order in nature and argued that the ancient belief in spontaneous generation would leave the door open to contingency and chance and, as such, deny God's omnipotence.

It was, as he put it, 'the royal road to atheism', and put into question our own origins. As he proclaimed in *The Book of Nature*, 'For if the generation of things be so subject to chance, what prevents man from being thus as easily produced in the same manner?'

Guided by this conviction and despite a scientific career that lasted only a dozen years, Swammerdam proved to be one of the outstanding comparative anatomists of the seventeenth century. He demonstrated that insects were just as complex as larger creatures and that no example of spontaneous generation could resist investigation. His observations became the foundation of our modern knowledge of the structure, metamorphosis and classification of insects.

MARIA SYBILLA MERIAEN

Over de

VOORTTEELING en WONDERBAERLYKE

VERANDERINGEN

D E R

SURINAAMSCHE

INSECTEN,

Waar in de Surinaamfche RUPSEN en WORMEN, met alle derzelver Veranderingen, naar het leeven afgebeelt en befchreeven worden; zynde elk geplaatft op dezelfde Gewaffen, Bloemen en Vruchten, daar ze op gevonden zyn: Beneffens de Befchryving dier Gewaffen. Waar in ook de wonderbare PADDEN, HAGEDISSEN, SLANGEN, SPINNEN en andere zeltzaame Gediertens worden vertoont en befchreeven. Alles in Amerika door den zelve M. S. MERIAEN naar het leeven en leevensgrootte gefchildert, en nu in 't Koper overgebragt.

About the Reproduction and Miraculous Changes of the Surinamese Insects by Maria Sibylla Merian.

The visual argument so clearly made by Swammerdam within this unchartered territory of metamorphosis was embraced early on by the acclaimed and unconventional German natural science illustrator Maria Sibylla Merian (1647–1717). Her research appears to have begun soon after Swammerdam's and, like Swammerdam, she consciously worked on the intersections of science and art, was drawn to spiritual ideas and was a brilliant artist and illustrator. The results of her work can been seen in the Library and Archives of the Natural History Museum, London (and other reputable organisations), including the images she drew and painted of the life cycles of 186 insect species in Suriname, in South America. Her meticulous watercolours depicting her careful observations of insects undergoing metamorphosis, particularly in their natural habitats, caught the attention of the Royal Academy more than 250 years before the first woman was permitted to join.

Natural History Museum Librarian Grace Touzel has studied Merian's legacy, which includes a superb collection of art. Not just any old art, but detailed works, including her creatively titled *About the Reproduction and Miraculous Changes of the Surinamese Insects*. In this mighty tome she highlights the life cycles of many insects and spiders, with many of her observations new to science, as well as being the first time these animals were drawn with their host plants. What is remarkable to me about this story is that, not only was this published in 1705 and that Merian herself went to Suriname in 1699, but her field assistant was her 21-year-old-daughter Dorothea – Merian was 52!

Touzel writes about Merian in *Rare Treasures from the Library of the Natural History Museum* (2017), stating that by the age of around 13, Merian was already rearing and studying silkworms. Born in Frankfurt in 1647, to a family of artists, Merian's biological father, Matthäus Merian the elder (1593–1650), was in her life for just three years before he died, ironically whilst bathing in Bad Schealbach (ne Langenschwalbach), which is known for its healing powers. But she was not without an inspirational father figure, for a year later her mother married Jacob Marrel (1613/1614–1681), a skilled still-life painter. And it was Marrel who taught and encouraged young Merian to develop her own artistic skills. She grew up fascinated by insects (I know how she felt) and by 11 was engraving copper plates. At 13 she was observing and making notes in her *Studienbuch* (Book of Studies). Just before her natural father's death, he had engraved the first volume of Johannes Johnston's *Historia Naturalis*, a truly remarkable book featuring all sorts of creatures, wonderous animals from across the globe – mostly real but with some genuinely bizarre and made-up. One of these volumes was on insects – the subtitle translates as '*Book of terrible insects with feet and wings*'. I have leafed through the spectacularly diverse pages of this book and can attest to its awe-inspiring drawings.

Merian wanted to draw and study all that she saw around her, and was lucky enough to have been given some silkworms, the respectable butterfly, as a gift. I say respectable, for at this time caterpillars (and worms) were still thought by most to 'generate' in mud and

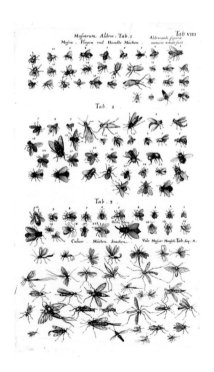

Illustrations by Merian's father in Johannes Johnston's *Historia Naturalis* (right).

Johannes Johnston's *Historia Naturalis*, Vol.3 – the book on terrible insects! (opposite).

HISTORIÆ NATVRA-
lis de Insectis. Libri. III.
de Serpentibus et Draconib, Libri II
Cum æneis Figuris
IOHANNES IONSTO
nus Med. Doctor
Concinnavit
FRANCOFURTI
ad Mœnum
Impensis
Hæredum Merianorū
M D C   L III.

M. Merian Iun. Inuentor

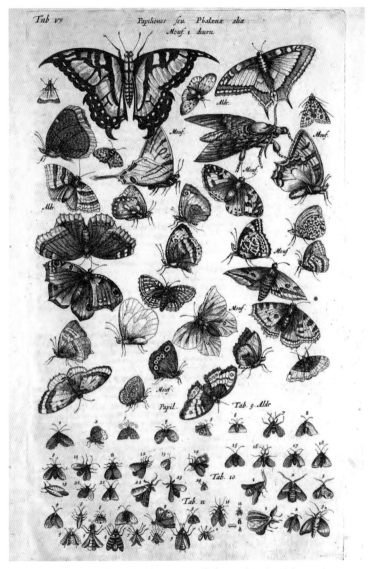

The terrible insects as drawn by the Merians (father and son) in Johannes's *Historia Naturalis*.

faeces, and so were not respectable. The adults were often thought to be alternative forms of witches! Quite specific witches – ones that curdle cream or steal your butter. The English name 'butter-flies' alludes to this devious past.

Merian studied this species and many more in her *Studienbuch*, in which she wrote about and drew her observations – today this consists of 285 drawings (some are missing) with many notes made on the back. Compiled over three decades, it is testament to her life-long passion. It was purchased from the Merian Estate by Robert Karlovic Areskin then, upon his death in 1719, acquired by Peter I, Tsar of Russia, and then with his death to the Russian Academy of Sciences.

Merian married the painter and publisher Johann Andreas Graff, her stepfather's favourite student, and moved to Nuremberg in Germany. Her husband was not the most successful of artists and to help with

The metamorphosis of a silkworm from 13-year-old Merian's *Studienbuch* or 'Book of Studies'.

as a life cycle strategy. It eliminates competition between the young and the old, as in many instances the larvae and adults occupy very different ecological niches. It also enables extreme adaptations of one or more parts of the life cycle for a particular role such as dispersal. Both adults and larvae can disperse – the larvae often thanks to an unwilling host!

Interrogating the individual parts of the life cycle is now taking on a new significance as an accurate tool to track environmental change across the globe. Chris Hassell is Professor of Biology at the University of Leeds, UK, and one of the many phenologists (the study of the timing of natural events over the years in relation to weather and climate) to study this. He draws in part on the UK Spring Index, used by the UK Government, which looks at four events to mark the advent of spring – the first flowering of hawthorn *Crataegus monogyna* (Jacquin, 1775) and horse chestnut *Aesculus hippocastanum* (Linnaeus, 1753), the first recorded flight of the orange-tip butterfly *Anthocharis cardamines* (Linnaeus, 1758) and the first sighting of a swallow *Hirundo rustica* (Linnaeus, 1758). Hassall has found that the orange-tip is now emerging eight days earlier than just 15 years ago. Insect population numbers, a change in a species' distribution, even a change in insect size are all indicators of a changing environment. But, says Hassell, choosing a model species whose life cycle is highly sensitive to temperature could be a biological proxy to tease out subtle changes in our climate.

Back in 2015, Hassell published a paper using dragonflies (Odonata) as 'macroecological barometers' to understand climate change. This is more important than just looking at weather data as it really does show us how nature is responding to our changing climate. He initially studied a dataset of more than four million recorded UK species (across 24 orders) but then honed in on just the insects, as they are better recorded and respond strongly to change. His favourite – the common blue-tailed damselfly *Ischnura elegans* (Vander Linden, 1820) – goes through one generation in two years in the northern regions of the UK, while there may be up to four in its southern range in the Mediterranean basin. He describes it as 'bomb proof' and shows a flexibility in both range, habitat (it doesn't seem to mind city living) and as such would make a great model in understanding climate change. Hassell is not alone in using

Blue-tailed damselfy – nature's barometer.

insects to predict the changes caused by climate change – across the globe many different species are being used. Increasing evidence is accumulating about how insects are responding to human-induced climate changes by shifting not only their timings and reproductive output but also range and diet. Dragonflies are great in that there is a wealth of citizen science recorders who can help us monitor these populations and help us model how the environment is changing.

Studying the life cycles of insects helps us understand our global environment, and what is equally important is that so many people can become involved in doing so. From the early insect aficionados, the determined and intrepid investigators, the studiers of form, function and habits, we are now in a position to understand how our environment is changing. From the most intricate of drawing to the simple snapping of a photo on a smartphone, the data is all important.

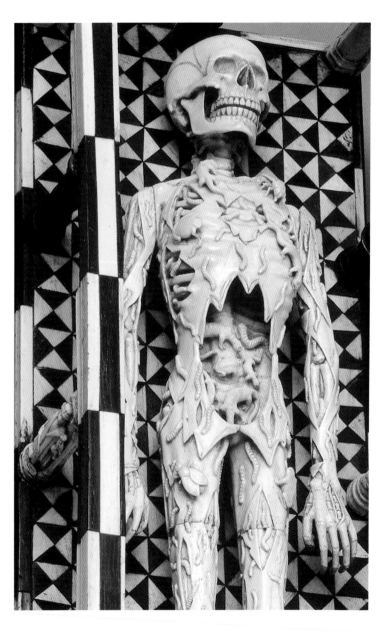

CHAPTER FIVE

Blowfly detectives

There was an old lady who swallowed a fly
I don't know why she swallowed a fly – perhaps she'll die!

Unknown

WHOEVER WROTE the song of the old lady eating a fly – and subsequently an entire menagerie – must have had a lingering anxiety of insects living inside us. This fear of insects and their association with dead bodies has been a long one. In the last chapter, we discussed the idea that insects spontaneously erupted from dead bodies, a theory that took science several millennia to quash. Religion has played a major part in our aversion, telling us that certain insects were associated with sin, and with those destined for hell. And art has depicted flies (and other less loved creatures) with human corpses for hundreds of years.

The earliest story of flies being associated with dead bodies is from Mesopotamian tablets 3,600 years ago. The Gilgamesh tablet XI contains the flood myth, where Gilgamesh, an ancient Mesopotamian hero, went to meet another hero, Utnapishtim, the 'flood hero'. In a story akin to Noah and his ark, Utnapishtim sacrifices several animals to appease the gods and halt a flood, after which the 'gods smelled the sweet odour of the sacrificial animal and gathered like flies over the

Staring death in the face. The animals of decomposition on *Memento Mori* attributed to Parisian sculptor Chicart Bailly.

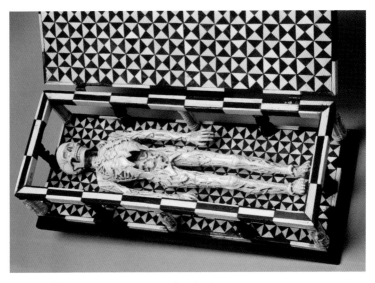

Chicart Bailly's *Memento Mori,* where flies feed on a corpse.

sacrifice'. The Ancient Egyptians depicted necrophagous animals, those species that feed on dead flesh that they realized were causing problems for embalmers, in their *Book of the Dead.* A personal favourite ode to flies and the dead is a coffin and cadaver made from ivory, attributed to the Parisian sculptor Chicart Bailly (*c.*1500 – *c.*1530). Bailly, like many Renaissance artists who sought to address the human concern of acknowledging death, while striving to create a personal legacy that might outlast it, left us a reminder – that we must all die and that other species will then inhabit us. The beautiful carving, titled *Memento Mori,* shows a body in an advanced stage of decay, being feasted on by flies.

As every good entomologist should have on their desk, I have jars of maggots, the larval stage of flies. All sorts of different types but specifically I have some from the family Calliphoridae, what are commonly called blowflies, blue- or greenbottles. Few people actively like these flies, because of their feeding habits – they are some of the very first insects to be attracted to, and feed upon, decomposing remains. In fact, the name blowfly comes from an old English expression 'flyblown' describing meat on which fly eggs have been laid. Even Shakespeare reported on them, in *Love's Labour's Lost* – 'these summer flies have blown me full of maggot ostentation'.

Calliphorids comprise of just over 1,900 described species. They undergo complete metamorphosis (as do all flies) and can be some of the shiniest components of our garden fauna, as well as some of the hairiest! For the female of these species to become a mother, she needs a massive amount of protein, and she is not alone – many species of Diptera are what we call anautogenous, whereupon they lack sufficient nutritional reserves when hatching as an adult, to start egg production straight away. Many, but not all species of mosquito for instance, require a blood meal before they can start a family. A paper published back in 2002 by Professor Richard Wall, University of Bristol, UK, and collaborators, describes an experiment whereby female adult calliphorids *Lucilia sericata* (Meigen, 1826) were restrained in modified pipettes. The end of each pipette had been cut off to give a wider hole through which a fly could pop its head out, and feed on different quantities of pig's liver. After a period of time, their ovaries were studied for egg development. Only the highest feeding regime (two meals of 10–20µl) 'allowed more than 50% of the flies to initiate yolk deposition and at least two meals of 27.5µl were required to allow maturation of eggs'. That is just 0.027 ml, which may not seem a lot to us, but it is critical to these flies, to ensure the development of the next generation. And because it is critical, they've evolved to locate food from many metres away or from what might be referred to as 'cryptic environments'. And with species such as the calliphorids that have a more ephemeral or unpredictable source of larva habitat, all that decomposing material, the adult flies really have to hone their food-locating skills.

Calliphorids' keen sense of smell would make a bloodhound blush with inadequacy, enabling them to seek out fresh blood and newly decaying matter in the most precarious of environments. It is their extraordinary smell skills that mark the adult blowflies out as useful early detectives. Calliphorids often live and breed in and around decaying flesh, and due to these life choices they've become major players in the highly specialist discipline known as forensic

A calliphorid and its eggs.

entomology – investigating insects recovered from crime scenes and corpses. But while flies may be fast and accurate in finding bodies, how does this translate into forensic use?

Dr Martin Hall, now predominantly retired (entomologists don't really retire completely), has worked in the Diptera section at the Natural History Museum since 1989, studying both forensic and veterinary entomology. During his career (and still after official retirement) Martin has acted as a consultant for many a police department or legal body and has worked on more than 200 cases. Nearly every time he becomes involved in a case, the first question he's asked is – how long has this person been dead?

Martin Hall states, along with every other forensic entomologists, that while he can't tell them exactly that, what he can tell them is when the flies first found the body. And why is that important? Well pathologists struggle to age a body after about three days from death (depending on the surrounding temperature). While insights from rigor mortis and blood pooling can help, beyond three days those clues become poor indicators. And that's when the flies can help out – a blowfly usually finds a body within minutes if not a few hours after death, whereupon it lays its eggs. So, these initial colonizers and their first generation start a biological clock ticking that more or less measures the time since death. A clock that has been studied and refined through years of research. That clock will keep ticking for days, weeks, months, and it's Martin's job to work out just how long that clock has been ticking for.

Solving a crime using insect evidence isn't new. It was first documented centuries ago in medieval China, when in 1247 the Chinese lawyer Sung Tz'u (or Song Ci or Sung Ts'u or 宋慈) wrote a training manual on criminal investigations called the *Collected Cases of Injustice Rectified* or more lyrically *The Washing Away of Wrongs*. And in doing so he became the first published forensic entomologist. Tz'u was a judge in the high courts of the Song Dynasty, and during his time in post he personally attended crime scenes. It is these cases, and the bodies examined, that were detailed in his book.

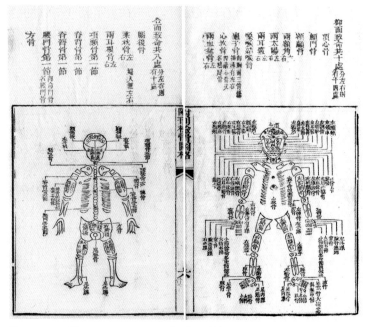

The naming of bones at crime scenes by Sung Tz'u.

In one case, he recounts the story of a murder near a rice field, back in 1235. The victim had been slashed (the direct translation is stronger in its language) most likely with a hand sickle, a tool commonly used for the rice harvest. But how could the murderer be identified, when so many workers carried these implements? Brian Knight, who translated the book into English writes, 'The local magistrate began the investigation by calling all the local peasants who could be suspects into the village square. Each was to carry their hand sickles to the town square with them. Once assembled, the magistrate ordered the ten-or-so suspects to place their hand sickles on the ground in front of them and then step back a few yards.'

It was, as Sung Tz'u recalls, a warm day, and before long bright green metallic flies began focussing on just one of the tools. 'The owner of the tool became very nervous, and it was only a few more moments before all those in the village knew who the murderer was. With head hung in shame and pleading for mercy, the magistrate led the murderer away.' The incriminating blowflies had been attracted to the blood and soft tissue stuck to the hand sickle. 'The knowledge of the village magistrate as to a specific insect group's behaviour regarding their attraction to dead human tissue was the key to solving this violent act, and justice was served in ancient China.'

It's an amazing account, and indeed *The Washing Away of Wrongs* forms a valuable manual for aspiring criminologists on how to go about investigating crimes. Martin Hall highlights one of the passages that struck a chord with him as a trainee investigator, 'don't be deterred by the smell of a corpse in the distance, but get directly involved yourself'. I liked the advice 'not to hide behind curtains' (i.e. get stuck in). Other cases solved by flies were to follow – 'A merchant was killed, and his silk was robbed in the road. A retired policeman was assigned to the investigation. After two days the investigator saw the boat on which there were plenty of flies clustered on the washed silk. Then the police arrested the men of the boat. Those men had to own up because the silk had traces of blood.'

Given this historic relationship with flies, it is perhaps surprising that Europe didn't cotton on how to use flies at a crime scene until the nineteenth century. Even more surprising given that European painters and sculptors had already observed this intimate if disagreeable relationship in their depictions of decomposing human bodies and their associated maggots. Documents from the Middle Ages, for example, illustrate maggots on corpses, including woodcuts with a 'dance of the dead' theme, whilst fifteenth-century oil paintings accurately depict the skull becoming skeletonized and internal organs shrinking, as a result of relentless maggot activity.

Carl Linnaeus (1707–1778) first described one of the more forensically important species of calliphorid in 1758, in the tenth edition of his ever-expanding work on biodiversity *Systema Naturae*. The fly was

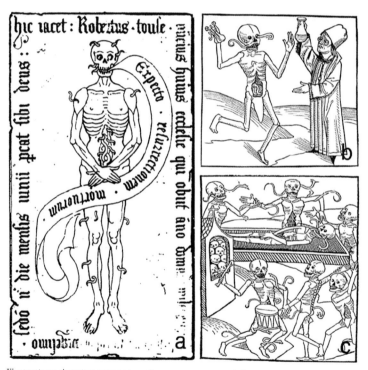

Illustrations showing maggots eating a corpse, especially eyes, nose, ears and mouth, where flies tend to lay their eggs.

identified as *Musca vomitoria* (now *Calliphora vomitoria*) and he had added a comment to his twelfth edition about their feeding habits, 'Tres Muscæ consumunt cadaver Equi, *æque* cito ac Leo', which translates as 'three flies consume the carcass of a horse as quickly as a lion'.

But it was in France and Germany, through the systematic exhumation of mass graves amidst a new wave of city planning and concerns for public hygiene, that doctors, examining at first-hand large numbers of corpses, began noting the insects on them. By the early 1800s, scientists knew that certain insects would inhabit decomposing bodies. Interest now turned to the matter of insect succession, the predictable order in which certain insects will visit a body. Physicians and legal investigators began looking into which insects would appear first on a cadaver, which came later, and what their different life cycles could reveal about time since death. In 1855, French hospital physician Louis François Étienne Bergeret (1814–1893), also known as Bergeret d'Arbois, was the first to take this new understanding of flies and

Vomito- 67. M. antennis plumatis pilofa, thorace nigro, abdomi-
ria.   ne cæruleo nitente. *Fn. fvec.* 1831. *
                          R r r f   *Scop.*

990   INSECTA DIPTERA. Mufca.

Scop. carn. 868. Mufca carnaria.
Gæd. inf. 1. *t.* 53.
Lift. gæd. f. 122.
Raj. inf. 27.
Reaum. inf. 4. *t.* 19. *f.* 8.
                     t. 24. *f.* 13, 15.
Geoffr. parif. 2. *p.* 524. *n.* 59.
Lyonet. lef. t. 1. *f.* 23, 27.
Habitat in Cadaveribus; *etiam* Americæ. *Kalm. Tres
     Mufcæ confumunt cadaver Equi, æque cito, ac
     Leo.*

Carl Linnaeus's comment in *Systema Naturae* about the voracious feeding capacity of *Musca vomitoria*.

cadavers and use it in the tragic case of an infant's death. The body was found by a French couple while they carried out some improvements on their new home. Suspicion fell instantly on the couple, even though they had only recently purchased the property.

Bergeret was knowledgeable about insect life cycles as well the colonization of a corpse by these animals. Using this knowledge, he was able to make an estimate of the time between the insects finding the body and the body being found by the couple, the postmortem interval (PMI). In 1855 he published a watershed report *Infanticide, momification naturelle du cadaver. Découverte du cadavre d'un enfant nouveau-né dans une cheminée où il s'était momifié. Détermination de l'époque de la naissance par la présence de nymphes et de larves d'insectes dans le cadavre, et par l'étude de leurs metamorphoses* ('Infanticide, natural mummification of the cadaver. Discovery of the corpse of a newborn child in a chimney where he had mummified himself. Determination of the time of birth by the presence of nymphs and larvae of insects in the corpse, and by the study of their metamorphoses'). This was the first time in modern forensic entomology that a PMI was included in the case report. This may be the actual time of death or not, depending on various factors, such as environmental temperature. Bergeret didn't get everything correct – he wasn't an entomologist, he had just read about it. He assumed that metamorphosis takes a whole year with the females laying eggs in the summer, which then transform into pupae (nymphs) the following spring, with the adults emerging in summer. Entomology may have played just a small part in the case – it was the mummification of the body that was the main focus – but it was still a tool to his calculations and helped convict the actual killers, which in this case were the previous tenants. They were successfully prosecuted.

Bergeret was an interesting character. He went on to publish *The Preventive Obstacle; or, Conjugal Onanism* in which he described why it was morally and physically wrong to have sex for any other reason apart from procreation. As a doctor he had seen many cases of disease and 'dysfunction' which he linked to this 'degenerative' behaviour. But in forensics at least he was valid in his belief that entomology was important

– apparently his last words were 'I wish to find out more about forensics'. Bergeret wasn't alone in his love of forensic entomology. Jean Pierre Mégnin (1828–1905), a French army veterinarian, produced *The Fauna of Cadavers* in 1894, in which he drew upon his 15 years in a Paris morgue to unveil revolutionary work examining successions of insects on corpses. By counting numbers of live and dead flies that developed every 15 days

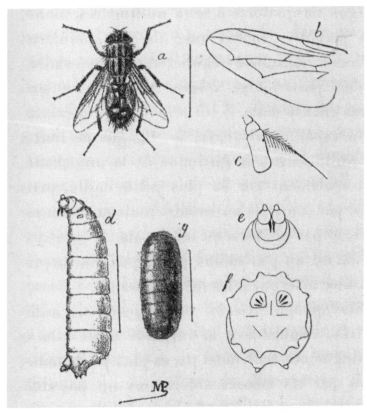

Images from Mégnin's *The Fauna of Cadavers*, the foundation of modern forensic entomology – the different stages of *Sarcophaga carnaria* (Linnaeus, 1758) (a, d and g) and external anatomy of larva mouthparts (e) and anal spiracles (f), as well as antenna (c) and wing (b) of adult.

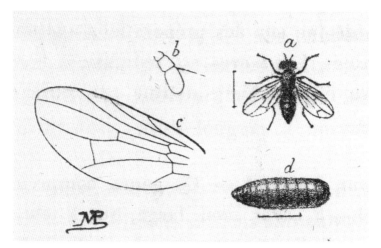

The adult, antennae, wing and pupal case of *Ophyra cadaveris* (Curtis, 1837) from Mégnin's *The Fauna of Cadavers* is now a synonym of *Hydrotaea capensis* (Wiedemann, 1818).

and comparing these with his initial count on the corpse, Mégnin was able to estimate how long they had been dead. He concluded that exposed corpses are subject to eight waves of insect successions, and just two if they have been buried. The publication of *The Fauna of Cadavers* firmly established the modern era of forensic entomology.

Today forensic entomology is able to draw on the work of a host of pioneer researchers, law enforcers and legal representatives, particularly from the past century, who've fought to legitimize applying entomology to help resolve crimes. As it has evolved, it has been applied to accidental and untimely deaths, homicides, suicides as well as archaeology and palaeontology.

This research is ongoing. Dr Gail Anderson became Canada's first full-time forensic entomologist in the 1990s, and is director of criminology at Simon Fraser University in British Columbia. She and her students and colleagues are active in full-time research, but she also assists with many criminal investigations. One of the more explosive cases she has worked on was not concluded by the presence of forensic evidence, but the complete lack of it. The accused was Kirstin Lobato, found guilty of a brutal murder in 2001, involving the mutilation of the victims' genitalia, and jailed for first degree murder for between 40 and 100 years. The story is convoluted, full of what appears to be much judicial error, but Anderson's key observation in the re-trial was the total lack of maggots on the victim. Maggots we know are deposited on an exposed body quickly. The adult females, if the conditions are right, can turn up within minutes, maybe seconds, and colonize the cadaver. The exact timings are influenced by temperature, season, amount of clothing, accessibility and time of day. Calliphorids are like us; they are diurnal, and they cease their activity at night (OK so not quite like us!). The body was discovered at 10pm with no evidence of maggots, which meant no flies, which meant the murder had to have occurred at night as well, a time during which Lobato had an air-tight alibi. Lobato was released after 16 years in prison, thanks to this forensic insight.

Anderson highlights that through forensic entomology we are also able to work out if a body has been moved, since it may have improbable fly fauna given its location. And if the site of the wound is no longer visible, we can use insects to determine where it was. Nowadays we can also use DNA extraction procedures to determine exactly who the maggots were feeding on. The actual details of handling entomological data are of course pretty complicated – it's not, as many think, a case of just turning up and studying the insects *per se.*

To arrive at any near accurate estimate of the time between death and discovery, a myriad of factors needs to be determined – the ambient temperature, the amount of clothing, extent of injury – all influencing the flies life cycle and that ticking biological clock. If the body is buried it is harder for the flies to access this and so delays insect colonization. The first test for British forensic entomology was back in 1935, in a landmark case featuring Dr Buck Ruxton, aka the savage surgeon. He almost got away with a brutal double killing known as the jigsaw murders, only to be thwarted by, amongst other things, flies. Ruxton killed his wife and his wife's maid and, considering himself aware of all things medical, he set about removing the forensic evidence. After killing his victims, he dismembered their bodies – hence jigsaw murders – as well as removing identifying features, which included a tattoo from the maid's leg. He then drove up to Moffat in Scotland, where he dropped the body parts off a bridge into a ravine. The visit Moffat website states that 'lying less than a mile from the A74(M), it's the ideal place to stop' – I am not sure they meant for people to stop and do this though. Two women were walking from the town shortly after and, when looking over Gardenholm Linn bridge, spotted what looked like a partially decomposed arm sticking out of a package below. The police were called and soon discovered more remains in an advanced state of decomposition.

The maggots collected from Moffat now sit in the Natural History Museum, among thousands and thousands of jars of preserved animals. In the Cocoon, as you go up the floors, past smaller and smaller specimens, you get to the top, to the insect collection. Within this there is one special jar, and I would argue one of the most important in terms of historical importance. It contains some very special maggots – the maggots removed from Ruxton's victims.

Maggots from Ruxton's victims, 1935, in the collection at the Natural History Museum, London.

There was no shortage of maggots on the body parts, and they were sent to Scottish physician, public health expert and forensic entomologist Alexander Gow Mearns (1903–1968). He was working at the University of Glasgow and had been studying how to determine the age of flies based on the length of their larvae. This knowledge enabled Mearns to estimate when the body parts were dumped. It coincided with Ruxton being in the area (accompanied by his son in the car!). The maggots were not used in court but provided preliminary evidence to help establish a timeline for events. The evidence against Ruxton was comprehensive and upon being convicted he was was subsequently hanged for his nefarious deeds.

Since then, many experiments have been conducted in laboratories and field sites around the world in a bid to learn more about the part insects play in the decomposition process – and in the most bizarre of crime scenes. Entomologists, including Anderson in Canada, are on a mission to further increase the robustness of that crucial PMI, the time between the first fly contact and the body being found, regardless of where it has been left, either out in the open or concealed in a car boot. To investigate bodies left in car boots, Anderson has what she calls Research Forests, in Vancouver. Carcases are placed in vehicles deep within these forests. These are not human carcasses as, understandably, there are restrictions on this, but carcasses of pigs. Pigs have been used for forensic studies for a while as they are of equivalent size and shape to humans, and they decompose in a similar fashion. There are invaluable research facilities where decomposing humans (and other animals) are indeed studied, colloquially called body farms. These were established back in the late 1980s in Tennessee in the USA, with similar facilities across America, in Canada and Australia. There are none in the UK as yet.

Back to the pigs. The bodies are placed in the boots of vehicles, as this is apparently a common method of human body disposal in Canada, and after several days Anderson examines them for maggots. Many different ambient factors need to be recorded to understand development, but once the maggots get into these concealed bodies they develop much more rapidly than those outside the boot – well it is a cosy, warm larder for them. In Anderson's experiments, this was the case with all but one of the cars. In this one exception, decomposition of the body was really slow, and they didn't know why, until they discovered there was a protective firewall that separated the boot from the main section of the car. So again, a scenario like this is a lesson that care must be taken in noting all the factors that may influence maggot life cycles. And it is not just cars that are investigated – a body has also been investigated that had turned up in a dishwasher.

Even some of the most secure spaces can be penetrated by a blowfly. Dr Hall at the Natural History Museum in London and his student Poulomi Bhadra, worked with colleagues at the UK's former Forensic Science Service on a couple of cases where bodies have turned up inside suitcases. We can guess a body in the open will have flies on it within minutes, but in a suitcase? Controlled experiments showed that while flies lay their eggs around the suitcase, they also insert their ovipositor – the telescopic ariel at the back end of the body through which the females lay their eggs – through the cracks in the zips and lay their eggs directly into it. Although there may be a bit of a delay between death and eggs being laid, it's relatively insignificant, and the flies are still able to colonize the body, even if at first glance it appears to be sealed off.

The majority of these investigations have focused on finding the larval stage of the life cycle, but what about the part after that, the pupal stage, before the new generation of adult flies emerge? Until recently, for any forensic investigation, this inability to know what's going on has been a big hindrance. But we can now age pupae, where previously we just knew how long it took between egg and squirming larva. It's an important breakthrough since it can account for more than half of a blowfly's life cycle. During this period, the opaque, barrel-shaped puparium houses a near-total body change. To us it is just a brown puparium – a seemingly inert stage showing no change until the adult emerges – but we now know that so much is happening in terms of reorganisation of tissues.

Computerized tomography (CT) scanning uses X-rays and a computer to create detailed images of living organisms (and dead ones, but it is the living ones that are important here). In much the same way that we scan living humans, we can use a smaller version – a micro-CT scanner – to build a 3D picture of inside a pupa, as its adult parts start to appear.

Dr Daniel Martin Vega is one of the researchers pioneering this innovative technology. Back in 2017, Vega, a research associate of the Natural History Museum and collaborator with Hall, published a paper visualizing what he referred to as the 'dance of life', where they were able to see what was happening during the relatively long pupation of a blowfly *Calliphora vicina* (Robineau-Desvoidy, 1830). They took X-ray images from every plane of view, then reconstructed them to obtain a visual sample that could be virtually dissected from any angle. Crucially, this scanning method increases the ageing accuracy of a

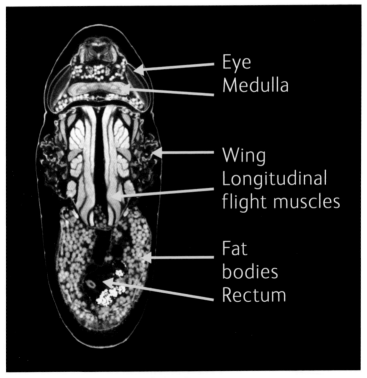

Eye
Medulla

Wing
Longitudinal
flight muscles

Fat
bodies
Rectum

Dorsal micro-CT scanner image of a blowfly pupa.

puparium and so helps to complete the forensic timing of an insect's arrival at the body. For example, you can see that in the pupa the larva's digestive tract undergoes a pattern of change, from a closed-inflated sac to a long and convoluted canal. These differences in morphology can be assigned to differences in the time of development, and so we are now able to determine accurately how old the pupa is. It's proving a promising method for providing accurate and reliable age estimations of blowfly samples.

The novel use of technology like this helps to boost entomological applications. But the development of our knowledge of forensic entomology has happened in stops and starts rather than a gradual increase. In fact, that can be said for all our entomological knowledge. There's much we still don't know.

Anderson remembers enquiries from the police at the start of her career in forensics, where they would call up to ask wearily if anything could be done with the 'insect evidence' they'd acquired. They just didn't see it as science. It's a significant contrast to now, when, thanks in part to mass media we are all aware of an insect's capacity to play mini-detective. Analysis and the expert opinion of forensic entomologists are now routinely solicited in both criminal and civil investigations. While there's much to learn, the field has potential to yield insights beyond that crucial 'time since death', as it continues to become an increasingly important tool within the forensic sciences.

So next time you sit in your garden with a glass in your hand, absently staring at an insect as it bursts out of its pupal home, take a moment to ponder it – and maybe you could be the next person to think about a novel development brought about by these marvellous beasts.

CHAPTER SIX

Dazzling disguise

'A gorgeous butterfly rushes out of the gloom into the shade and is in a moment seen to be a painted novelty; then comes the excitement of pursuit; the disappointment of seeing this beacon dance over a thicket just out of sight; the joy of finding it reappear.'

Margaret Fountaine (1893)

THE IRIDESCENCE of insects including those seen on a butterfly's wing is one of life's joys, a joy that lives on years after the insect itself has died. Among the collection at the Natural History Museum there are drawers of butterflies with wings so vibrant they appear to have been collected yesterday. But many of these are decades old, bearing an iridescence so unnatural, durable and bright you wonder how they've maintained their youthful countenance. Biologists have long debated the exact purpose of these mesmerizing colours and why they are so widespread in nature – camouflage, attractant or conspicuous warning? And could modern technology ever replicate such vibrancy to produce environmentally friendly and durable artificial colours?

Few of us can resist the draw of many brightly coloured butterfly species as they flit between hedgerows or flower beds, either dining on nectar or trying to satisfy their urgent need to attract a mate. And you wouldn't be alone in your love of the Lepidoptera. Of all of the insects it

One of the many morpho butterflies in the Natural History Museum's collection. This one is nearly 100 years old and still looking good.

is the 'butterflies' that have been collected the most. I use quote marks for the term butterflies as, with many of the insects, there is confusion as to what they are by many a lay person. Butterflies and moths make up the order Lepidoptera – a term first used by the father of classification Carl Linnaeus in his 1746 *Fauna Svecica*, and derived from the Greek *Lepidos* and *ptera* meaning 'scale wing'. All the insect order names are Greek. My mother studied Greek and Latin and as I once read out names to her, she was able to guess what insect I was referring to – it was all Greek to her, but she understood it! Within the order there are more than 140 families and, as with all thing's insect, taxonomists are still trying to organize them. But most, nearly 99% of the 180,000 described species, belong to the clade or group Ditrysia, named after the female's genitalia. Now there's a sentence I don't often write! These females are thought unique in their genital arrangements in that they have two separate sexual openings, one for mating and one for laying eggs, hence their name di meaning 'two'. It's within this group that the butterflies are found.

Butterfly or moth? Surprisingly these are all moths.

114

Your knowledge of what a butterfly is might stem from what you were told as a child – that butterflies are colourful and fly during the day, moths are drab and active at night, that butterflies hold their wings closely upright over their backs at rest, and moths splay them flat over their back. The advanced among you might also think that moths have club-like antennae and butterflies don't. Turns out this isn't the case for a lot of them! For example, originally there were two superfamilies Calliduloidea (Old World butterfly-moths) and Hedyloidea, containing only one family Hedylidae (New World butterfly-moths or moth-butterflies). Both superfamilies contained many species that looked like both moths and butterflies with our simple classifications, but only Hedyloidea contains true butterflies. A third superfamily Papilionoidea in fact contains the true butterflies (and now the Hedylidae have joined them). (Really, they should all be called moths as indeed this superfamily nestles within the 'moth' clades, but I guess the public may have some issues with that). Papilionoidea comprises more than 18,500 species

A butterfly species now extinct – *Papilio sloanus (= Urania sloanus)* (Cramer, 1779).

within seven families: Papilionidae (swallowtails), Hedylidae (nocturnal butterflies), Hesperiidae (the skippers), Pieriidae (whites), Riodinidae (metalmarks), Lycaenidae (blues) and Nymphalidae (brush-footed). It's these families that have captivated scientists and naturalists for centuries.

It was the Victorians that really caught the butterfly collecting bug. And significant among the throng of beard-touting males was the female

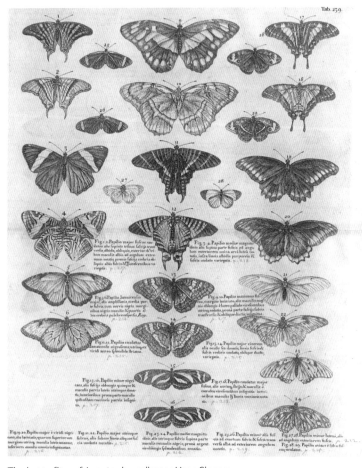

The butterflies of Jamaica by collector Hans Sloane.

adventuress and passionate entomologist Margaret Fountaine (1862–1940) who was drawn to all things butterfly – or as she said 'diurnal lepidoptera'. From her dedication, wealth, and intrepid travels would emerge one of the largest and most colourful butterfly collections in the country.

Fountaine was the second child and eldest daughter to Mary Isabella Lee and Reverend John Fountaine, and one of eight – John, Margaret, Rachel, Evelyn, Geraldine, Arthur, Constance and Florence, all of whom were very young when their father died in 1877. Margaret was just 16 when her father's death necessitated a move to Norwich. The date was 15 April 1878. She wrote about it in her diary. Diaries of what she did and saw as a butterfly collector in later years would also become her legacy. There would be many. She ends her entry for that date with 'and here ends the account of how I spent the 15th of April 1878, so good-bye for three hundred and sixty-four days' – it would be a little longer for anyone else to read them for their contents were to be kept secret for a hundred years from that date.

As a young girl in Norwich, Fountaine spent time sketching the cathedral, exploring botanical gardens, and visiting the butterfly collections of family friend Henry John Elwes (1846–1922). Elwes' collection was to eventually amass 30,000 butterfly specimens, now housed at the Natural History Museum, London. Her diaries at this stage chart the comings and goings of love. Her second diary entry was made a year from the first, a date that was Margaret's day – for all the children had a date separate to their birthdays that was their day – and was a confession of love. By 1883 she has fallen in love many times and writes that her doctor, Mr Muriel, has said that 'both Evelyn [sis] and I have got a monomania and that was the love of men'. Although she also writes about her love of painting, there was little in those early transcripts about what would become one of the real loves of her life.

Fountaine was not particularly fond of her mother whom she described as very strict and a scourge. Her mother by all accounts had quite a 'confined life', being a vicar's daughter and then marrying one and then, well, spending many years birthing! Her mother was herself one of many, with 13 siblings, as was her father, who was one of

10. Thankfully it was all those uncles and aunts who looked after the children financially after their father's passing – and two of them in particular were to be of great importance.

In June 1886, Fountaine's uncle Sir John Bennet Lawes (1841–1900) brought news of a windfall that bought her independence. Lawes had joined the family by marrying Caroline, Fountaine's paternal aunt, and was described as a scientifically minded landowner, what we think of now as an agricultural scientist. He was raised in Rothamsted, Hertfordshire, an English village that would become synonymous with agricultural science. The Rothamsted Experimental Station was born out of his research into plants, manure and then the more lucrative manufacturing of his own fertilizer. In 1842, Lawes patented the first artificial manure and built what is thought to be the first plant for artificial manure manufacture in Deptford, UK – if not the world. And then he raked the cash in! Luckily Lawes persuaded Edward Lee Warner, another uncle to Fountaine, to invest in his work and crucially to leave some of his new riches to his nieces. With that money Fountaine described herself as free, free to do anything she wanted, free to find her wings. In 1891, Fountaine and her sister Florence travelled for the first time from the UK to Switzerland, which is where her love for butterflies took off.

'I would often spend my afternoons at St Jean and go out with an English girl after butterflies, a pursuit which once started soon became all-absorbing. I filled my pocket box with butterflies, some I had only seen in pictures as a child and yet recognised the moment I caught sight of them on the wing. I little thought years ago, when I used to look with covetous eyes at the plates representing the Scarce Swallowtail or the Camberwell Beauty that I should see both these in a valley in Switzerland and know the delight of securing specimens. I was a born naturalist, though all these years for want of anything to excite it, it had lain dormant within me.' – Geneva, 1891.

For the next 50 years, Fountaine toured the world – astonishingly for the time, on her own – to follow her passions of art, music and butterflies, collecting specimens in 60 countries on six continents. Her stunning collection of more than 20,000 butterflies is housed at Norwich Castle Museum, part of the Norfolk Museums Service, where

Dr David Waterhouse was senior curator of natural history and geology. (As an aside, Waterhouse and his legendary predecessor Dr Tony Urwin became curatorial heroes when they tackled a thief trying to steal a rhino horn from the museum's display. Never mess with a curator!) Waterhouse also appears to be one of the many men who fell under the spell of Fontaine. Or maybe it's just her collection? For hers was one of the first to be acquired when the museum was formed in 1825 and it was vast – 22,000 butterflies were donated from across the globe that she had collected over nearly five decades. Waterhouse writes that Fountaine had wanted it to go to the Natural History Museum, London, but upon hearing that the collection would be broken up she bequeathed it to Norwich.

Margaret Fountaine's morpho butterflies.

One of Margaret Fountaine's twelve cloth-bound diaries.

On the day I visited, Waterhouse had some of the 'morphos out' because not only are these the most beautiful iridescent butterflies from South America, but they're his favourite. And indeed, some of them are the most perfect specimens you'll ever see. No battered or torn wings here, missing legs, or the dishevelled look of some insects caught on the wing. That's because Fountaine had worked out that, if she studied their life cycle, she could raise them intact. So, she often collected females and would rear their eggs – maybe looking after 100 or so caterpillars, selecting the most perfect of the adults for pinning, and releasing the rest (an early *ex situ* conservationist!).

Fountaine recorded all her observations and experiences in her diaries, which she took with her everywhere. These diaries, 12 in total, are also with her collection, all cloth-bound and sealed in a container. They reveal a woman who was not just a serious scientist, but one who also lived her life to the full, taking the place of the butterfly, rather than of the collector, when it came to relationships with men. There are accounts of bicycle trips through France, as well as motorcar excursions across Tenerife, which she undertakes with eight young Spaniards. She joined a gang of bandits before sailing onwards for Cuba and Chile. There was Mr Standen in Corsica, the father of six single daughters but with the spirits of a school boy when once out in the fields, net in hand. There was a Mr Champion, who collected beetles, and 'the brothers Jones – Fly Jones the slayer of butterflies and Paint Jones, whose watercolour sketches of Corsican scenery are as full of talent as he is of conceit, which is saying a great deal in favour of the sketches'. What would her mother have said? The accounts of her many adventures and encounters were sealed inside a chest with specific instructions to not be opened for 100 years. Can you imagine the excitement in 1978 when the time came to open these mysterious diaries.

Waterhouse, although not around for the opening has read them and recounts Fountaine's infatuation with Septimius Houston, the head

Margaret Fountaine collecting her precious butterflies.

chorister at Norwich Cathedral. As he puts it 'she stalked him' – she was obsessive, and for years this persisted, an obsessive personality that flitted between her interests in men, singing, travel and of course the natural world. Amusingly, it was on the road from Damascus station in 1901 that she had her road to Damascus moment. For it was then she met Khalil Neimy, the Syrian Dragoman interpreter and guide at the hotel she was staying at. She is most entertaining in her first descriptions of the man who was to spend many years as her hired hand and fixer, and one of her great loves – 'I soon saw he was the most awful liar, but I thought that he might be extremely useful'. Indeed, he was a man ardent for her hand in marriage, but who had already taken the hand of another. By the time she found out, she was already committed to him, fake engagement ring and all! And so, after a short hiatus, their relationship resumed. And across the globe they collected, during which she periodically returned to England to deal with her family, her diaries, and most importantly, her specimens.

While assembling her collection, and examining her various finds, Fountaine thought more about her butterflies, and was becoming alert to the way nature was acting as an artist. Unlike pigments that absorbed and reflected certain wavelengths of light, the multiple scales on the morpho butterfly wing caused light to diffract, creating extraordinary shimmering optical effects and changes in hue, from full vibrancy to near invisibility, depending on your angle of inspection. Those scales that gave the order its name were the reason these species dazzled and remained dazzling for years after.

And it is here that the story takes on a fascinating dimension, for unlike the black margins of the morpho's wings, the iridescent blue is not a pigment. If these specimens had been left out in the sunshine the black pigment would have faded, but the blue wouldn't. So why is that? Well, the blues are a structural colour – that is they are in effect tiny bumps (sometimes known as nipples) on the surface. They were first observed and written about by Robert Hooke in his *Micrographia* of 1665 where his observation 36 is of 'Peacocks, Ducks and other feathers of changeable colours'. 'The parts of the Feathers of this glorious Bird appear, through the Microscope, no less gaudy than do the whole

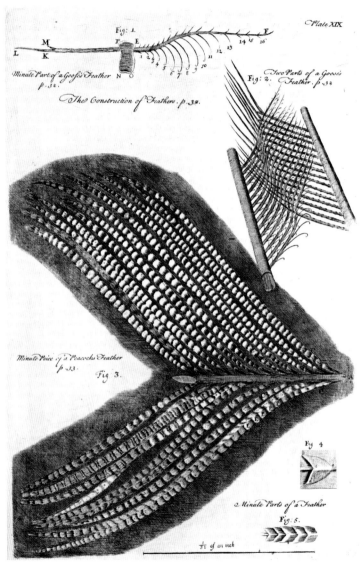

The feathers of a bird as drawn by Robert Hooke in *Micrographia*.

Feathers; for, as to the naked eye 'tis evident that the stem or quill of each feather in the tail sends out multitudes of Lateral branches… so each of those threads in the microscope appears a large long body, consisting of a multitude of bright reflecting parts.'

Hooke goes on to write that this structural colour very 'much diversifies the appearance' and that the colours are caused by 'reflection and refraction'. Although Isaac Newton (1642–1727) developed the idea that light is comprised of particles in 1704 – the particle theory of light, or as he called it 'the Corpuscular theory of light' – it was only in 1801 that Thomas Young (1773–1829) determined that light behaved as a wave, and that changes in the shape of the object altered the waves hitting them, creating interference patterns. Theories, as is their way, as to why species are patterned or coloured, abounded for the next 100 years and in his book *Animal Coloration* published in 1892 Frank Evers Beddard (1858–1925), an English zoologist, summarized all the knowledge on the subject in an easy-to-understand way. A fabulous review of this publication the following year highlights that where many of the 'superficial writers and the … still less discerning public' are keen to believe in theories without question, Beddard however is not, and interrogates the function of both pigments and mechanical colours of animals in terms of sexual selection, mimicry and defence. Many of the species seen in museum collections are bold in their coloration, and bright colours have often been thought to be an evolutionary trade off – they can attract a mate, yet they might also attract a hungry predator. But there was something else.

In the late nineteenth century, the celebrated American artist, naturalist and freethinker Abbot Handerson Thayer (1849–1921), proposed that these showy splashes of iridescent light could, in a natural setting, act as a highly effective form of concealment. Thayer had built a successful career painting everything around him, from portraits of his own children to the animals and landscapes that he passionately observed. He was also a pre-eminent technician in paint, and an acknowledged American master of the colour theories developed in Munich and Paris that explored how colours have a different hue and chroma, different intensities, and of how colours when juxtaposed together enhance or cancel one another out.

Driven by his obsession with both art and nature, Thayer published a six-page article titled *The Law Which Underlies Protective Coloration*, in which he intended 'to set forth a beautiful law of nature which, so far as I can discover, has never been pointed out in print. It is the law of gradation in the colouring of animals, and is responsible for most of the phenomena of protective coloration except those properly called mimicry.' This protective or defensive coloration in which animals (or plants) appear brightly coloured but are, through that colour, concealed, was a concept that in Thayer's mind hid something even more wonderful. Thayer had realized that animals could 'disappear; and that 'brilliantly changeable or metallic colours are among the strongest factors in animals' concealment'. He firmly believed that iridescence should be categorized as a camouflage strategy across the animal kingdom. But how can colours that are both brilliant and changeable contribute to concealment? This almost counterintuitive effect was achieved, he claimed, by the hues of light appearing to flatten an animal's contours and distort its tell-tale shape – in effect, hiding it in plain sight. In 1909 his newly discovered law (who doesn't want to discover a law?) stated that animals don't try to hide by looking like something else, a twig for instance, but 'cease to exist at all'. The drama of it!

Thayer's law consisted of two principles. The first was obliterative countershading, whereby the contrast between an animal's shadowed undersides, and its upper areas are equalized, thereby concealing its self-shadowing. The second was disruptive patterning, in which strong arbitrary patterns of colour break up outlines, so the animal either disappears or its shape is distorted. He illustrates this with some crude diagrams and a series of photos of dead birds positioned in scrub at different angles. He encourages the reader to try this out for themselves at home – not sure how I could explain what I was doing to my neighbours. Later in the same year came a second publication. This time Thayer carved woodcock-sized eggs (44 x 31 mm/1¾ x 1¼ in) and painted them in the colours of grouse and hares and placed them in the undergrowth. Now comes my favourite part. He then 'summoned a naturalist' who subsequently failed to find the colour-graded eggs even 'when told

Hidden in plain sight, Thayer's ducks. The visible one (left) has not been countershaded, whilst the duck (right) is countershaded and all but invisible.

where to look'. He even painted bright spots on some of them, which just appeared to the naturalist to be part of the landscape.

Thayer held big public demonstrations of his observations throughout America and Europe, and within the Hope Museum of Natural History, Oxford, the Cambridge Zoology Museum and the Natural History Museum, London. Various models were used to highlight the effect. In November 1896, he arrived at Harvard Museum of Comparative Zoology, USA, and delivered his new law through the medium of sweet potatoes! His views added to a debate on the coloration of animals that had raged since the publication of Darwin's *On the Origin of Species* in 1859. Darwin and his co-theorist Alfred Russel Wallace differed in their views on why animals were coloured, but Thayer was able to lend a valuable painterly eye combined with his naturalist knowledge to the subject. He was helped enormously by the invention of 'instant photos'. We barely give these a second thought in this age of instant media, of having small, powerful cameras we can carry in our pockets, but this new technology replaced earlier cameras that were large, cumbersome objects that required careful handling of chemicals to produce the images, and took time. With instant photos, Thayer was able to take images of animals in the field and quickly see them – or rather not – in their natural environment.

This new law by Thayer, where the colour on an animal can help it 'cease to exist' received lots of attention. As his son Gerald notes in the preface to their 1909 joint book publication *Concealing-Coloration in the Animal Kingdom,* it was well received especially in England. Thayer went as far as creating a coat of painted rags that disrupted the outline of a human, and wrote to the US navy suggesting this kind of camouflage could be useful for ships but failed to convince them. (On a side note, Norman Wilkinson, a British marine painter, in 1917 devised dazzle painting to protect the UK navy from the German U-boats.) But the theory still had its critics. Many scientists angrily challenged Thayer, correctly arguing that conspicuous colouring was also designed to be very much seen, to warn off a predator or attract a prospective mate. In particular, they resented Thayer's insistence that his theory applied to the entire animal kingdom. His most famous detractor was big-game-hunting US president Teddy Roosevelt, who publicly scoffed at Thayer's idea. From his own experience, Roosevelt knew that zebras and giraffes were clearly visible in the African grasslands, from miles away. 'If you … sincerely desire to get at the truth,' wrote Roosevelt, 'you would realize that your position is literally nonsensical'. It wasn't until 1940 that Thayer's law of obliterative countershading received an acceptance in print, when a prominent British naturalist, Hugh B Cott, published *Adaptive Coloration in Animals.* While criticizing Thayer's overenthusiastic attempts to explain all animal coloration as camouflage, Cott gave strong support in particular cases to this law of concealment.

There are still questions about Thayer's law, but many zoologists were quietly receptive to the idea. Dr Karin Kjernsmo, University of Bristol, UK, is a behavioural and evolutionary ecologist working in the CamoLab and studies how animals use colours and patterns to avoid being eaten by their enemies. She argues that the most convincing cases that support Thayer are when we find iridescence in the non-reproductive life stages of an animal. Iridescence is found in both the adult and juvenile form of many insects. The larval forms, such as the beetle grubs and the butterfly chrysalises, aren't sexually reproductive, and so they aren't trying to show off to the opposite sex with their fancy patterns and ornate colouring. The larval stage has just one thing to do, and that is eat while trying to avoid being eaten. In many ways the larvae are the most defenceless – many are inactive, so they have to disguise themselves, adopting the strategy that Thayer highlighted, 'The animal's surface, is, by its iridescence, made to appear 'dissolved' into many depths and distances'. Thayer seemed way ahead of his time, not only in suggesting how the best disguises are dazzling, but also in the way this structural colour is created. It has been more than a century since his theories were published, but evidence to support them has only happened recently, thanks again to advances in imaging, this time electron microscopy.

The most common form of iridescence seen, for example in beetles, is multilayer iridescence. The exoskeleton, as we learnt in chapter one, is made up of multiple layers of chitin. Kjernsmo explains that, as the white

The iridescence of an Asian jewel beetle, *Sternocera aequisignta*, can help it hide from predators.

light passes through each of these layers the differently-sized wavebands are either reflected or cancelled out depending on the spacing between these layers. Only certain wavelengths can be detected by us and all other animals. Sunlight is a complete mix of wavelengths, or colours, resulting in white light. But when this white light hits an object, it's split into these different wavelengths that are either absorbed or radiated. And so, it's these structures on the animal that define the different colours we can see. The colours we see in the beetle depend on the spacings within its layers of chitin.

Kjernsmo looked at the wing cases called elytra of the Asian jewel beetle *Sternocera aequisignta* (Saunders, 1866), to further explore why some animals have such bling coloration. Jewel beetles (Buprestidae) are aptly named and have been used as jewellery all around the world. Victorian ladies in the UK, thanks to their fashion whims, caused the death of millions of them. Gowns were

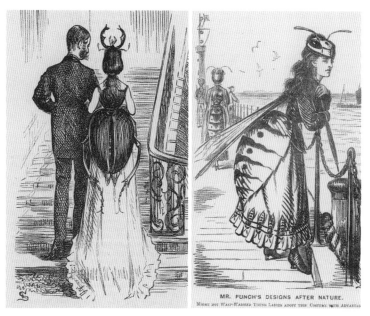

MR. PUNCH'S DESIGNS AFTER NATURE.
MIGHT NOT WASP-WAISTED YOUNG LADIES ADOPT THIS COSTUME WITH ADVANTAGE

Illustrations from *Punch* 2 April 1870 and 16 October 1869.

embroidered with the wing cases and so zealous were the ladies for these iridescent creatures that *Punch* caricatured many in its publications. In keeping with Thayer's experimental work, Kjernsmo positioned beetle elytra on different plants that ranged in colour, gloss and background. She first observed if birds went for the wing cases. She then 'summoned' once more a naturalist (or in this case probably a student), to try and spot them too. If neither could spot them, then it was camouflage. But if the humans spotted them, and yet the birds didn't go for them, then possibly it was 'distasteful' or aposematic coloration, where the birds were put off. But they didn't. Neither could spot them. What's more, the glossiness of the adjacent plant leaves further aided their concealment.

The counter-intuitive idea – that bright colours can deter predators – had been a fascination for Fountaine. She illustrated the life cycles of many colourful insects, now safely stored in the Rare Books Room at the Natural History Museum, curated by Special Collections Manager Andrea Hart. Fountaine was alert to the extraordinary optical effects this structural colour might have had in deterring predators, and she used all sorts of everyday objects to help her create the same effects on paper. Hart has found that for the chrysalis illustration Fountaine used the bright foil lining of a box of cigarettes. And as shown in Kjernsmo's field experiment, in the diagrams of the host plant drawn alongside the chrysalis, you can see how the colour and structure of the plant hides the static creature.

It's this unique, deceptive property of iridescence, along with some of the most amazing colours we know of in the natural world, that has drawn attention from far beyond entomologists. For more than three decades bioengineer Professor Andrew Parker, a visiting Research Fellow at Green Templeton College, University of Oxford, UK, has been trying to recreate the dazzling properties of morpho butterfly wing cells, to generate high impact hues for a host of different commercial products, such as paints that could be used to make cars that stay shiny forever, or clothes that never lose their colour. Parker heads a research team that studies photonic structures and eyes, and his 2006 book *Seven Deadly Colours: The Genius of Nature's Palette and How it Eluded Darwin* discusses how colours are produced in the natural world, and how we humans could copy this.

Parker's early attempts to recreate natural iridescence involved

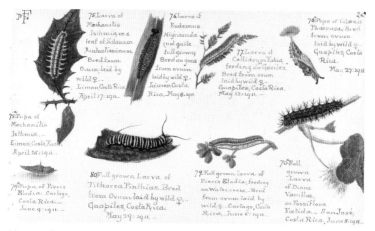

Margaret Fountaine's notebook showing the metallic pupal case of the *Mechanitis polymnia* (Linnaeus, 1758) (= *Mechanitis isthmia* Bates, 1863 is a synonym).

The metallic magic of metamorphosis. The chrysalis of *Mechanitis polymnia* – the orange-spotted tiger clearwing.

culturing butterfly tissue in the lab. He took the cells from a chrysalis that develop into wing scales in the hope that, with sufficient nutrients, each individual cell would produce thousands of scales. He would then be able to syphon off the scales and put them into a transparent liquid that could, for example, become paint, and the scales would provide the paint's iridescent colour. But during the process of converting cells to scales, part of the original cell was repeatedly lost, and the yields proved disappointingly low – one cell produced no more than a single scale!

More recently Parker has entered the realm of hi-tech mechanical engineering and founded Lifescaped, a company that makes extremely thin layered sheets of what he calls pure structural colour, resonant colours that maintain their intensity at all angles and which will not fade in the sun. He is guarded about the exact methodology, other than to say the precise mechanism is not found in nature but can be produced in large quantities using combinations of silicon dioxides. Nanoscale patterns created in the sheets scatter light into colours that are visible from all directions, and different hues can be produced by changing the dimensions of the pattern. The sheets can then be used to cover objects. In 2020 Parker teamed up with Nike to produce one, very expensive, pair of trainers. The coating is just 1/100,000th of a hair thick and it's the brightest of colours. They are what Parker aptly describes as mesmeric. This is butterfly technology on your feet!

It's not just for fashion reasons that folks are excited. There are much more practical and environmentally sensible reasons for developing structural colour. More recently Parker has been adapting industrial machines to paint objects. Think about painting a plane, say an Airbus A380, the largest passenger aircraft at the time of writing. It requires 3,600 l (792 gallons) of paint and each layer of 0.2 mm (0.008 in) adds an extra 650 kg (1,433 lbs) of weight. Each time you repaint, you add weight, which increases the amount of fuel needed to fly the plane, which is both costly and environmentally damaging. Using a 'paint' that never fades will save both pocket and planet.

And it's not just one colour that Parker can replicate, it can be many – he states that the technology can produces all the colours in the spectrum as well as all the finer shades – that's one massive colour

A seven-layer pigment created the iridescence of the Lexus Structural Blue model inspired by the structural wing colour of the blue morpho.

chart. The possibilities are limitless, and if we are able to figure out how to control colours from different angles of inspection, we may even develop invisible 'cloaking' devices, currently the stuff of science fiction. But a gentle word of caution here. Kjernsmo says iridescent packaging is already on our supermarket shelves. Not to draw us in, but possibly, as with the butterflies, to camouflage the list of ingredients, subtly concealing it from scrutiny. Sneaky…

CHAPTER SEVEN

The ultimate upcyclers

'In the north at certain seasons a small insect or midge comes in clouds, filling the air to a great height. The people gather them at night and bake a cake from them; they are called "Kungo", and the cake is said to taste not unlike caviar, or salted locusts'.

David Livingstone (1865)

I AM NOT NORMALLY allowed food at my desk, for fear of encouraging pests into the Natural History Museum collection. But I do have a piece of cake. It looks like a rock, and is composed of thousands upon thousands of flies, squashed together to form a patty. This is a kunga cake or kungu, made from gnats that form extensive swarms throughout the year over Lake Malawi (also called Lake Nyasa or Lago Miassa depending on which country you are paddling your toes in it from) – one of the African Great Lakes and the third largest and second deepest in Africa. As well as being home to over 700 species of cichlid fish species, many of which are bred for the aquarium industry, it is home to 28 species of freshwater snails, one species of crab and, among the many insects, one species of midge called *Chaoborus edulis*. These are specifically a species of phantom midge from the family Chaoboridae and, as their name *edulis* implies, these are edible to us humans.

David Livingstone (1813–1873), the intrepid and infamous Scottish explorer, wrote of this midge in his 1872 book *Livingstone's Africa: Perilous Adventures and Extensive Discoveries in the Interior of Africa,*

Fancy celebrating your birthday with a kunga cake?

but it appears that even though he may have been adventurous in his habits it didn't stretch to him actually trying the cake.

Midges may seem an odd ingredient to those with a western diet, but they make protein-rich burgers, patties, cakes or however you wish to consume them. And they probably have one of the easiest harvesting methods I have come across. You simply oil a frying pan or skillet and then wave it around the swarm, where the midges get themselves stuck to the pan, and can then be cooked. I'm no stranger to eating insects, and see them as kinds of land prawns. I've eaten baby bees that tasted like pockets of honey, I've eaten crickets and grasshoppers and I have even drunk caterpillar poo tea – yes, there is a speciality tea made from the faeces of the grain moth larvae that feed exclusively on *Camellia sinensis* (L.) Kuntze – the plant that gives us tea leaves. It is thought to have many advantages over your regular brew, having more nutrients, vitamins and proteins. To me however, it was not as satisfying as your regular cuppa – in fact I found it slightly dehydrating, the opposite of what you want! This is not the only insect tea. There are many varieties of insect that are fed many different plants to produce differing flavours. Drinking and eating insects is a global phenomenon, and they have always been a part of our diets. Entomophagy is the name given to insect eating, and the Food and Agriculture Organization of the United Nations (FAO) estimates that today more than two billion people supplement their diet with them.

Kieran Whittaker is the founder and CEO of Entocycle, an insect technology company set up to produce high-grade sustainable insect protein. Whittaker came up with the idea after spending five years as a diving instructor in some of the most beautiful places on the planet. From Thailand to Mexico he taught, travelled and healthily ate his way across the globe and in doing so he realized there was a problem. And that was we were consuming protein with a huge carbon footprint. In a report commissioned by a leading British supermarket, Tesco, but undertaken by the Oxford Martin School's Future for Food programme at the University of Oxford, it was found that the British consume 2.5 billion beef burgers a year. Add that to a roast beef dinner and many other beef products and that is the equivalent in CO_2 emissions of one person driving a car for 43.5 billion km (27 billion miles) or, with some beer mat statistics, every car owner on the planet driving 28,968 km (18,000 miles) a year. Add in the meat we consume from chickens, pigs, fish and everything else that we farm and you see the point. Before Whittaker was a diver, he had studied environmental design and so he put this to good use in 2017 by starting up a maggot factory underneath some railway arches in London. Maybe I do a disservice in calling it a maggot factory but essentially it is, albeit a very high tech one. And it might go a considerable way to helping save our planet.

When you enter the Entocycle offices it's not immediately obvious what happens there. But behind some heavy doors is an insectary, row upon row of containers, within which many male and female black soldier flies *Hermetia illucens* (Linnaeus, 1758) are doing what many males and females do when they get together. And these wasp-like black soldier flies produce the original hipsters – their larvae eat only organic and love to recycle. It is these glutenous larvae, or maggots, that are of particular interest.

The black soldier fly – *Hermetia illucens.*

The munching maggots of *Hermetia illucens*.

Armed with powerful chewing mouthparts the maggots have a remarkable ability to shred, devour and transform nearly any kind of organic waste into high-quality edible protein, all the while leaving a carbon footprint much smaller than their own little forms. The other great thing about these maggots is that the subsequent adults themselves don't consume the waste or soluble food that many other decomposing species do, and so don't spread any disease. Whittaker describes them as 'superflies' and they are indeed seen as the crown jewels of a fast-growing insect-farming industry, addressing a growing need to find cheap, clean reliable protein for livestock and pet feed, and possibly much more. As the global demand for protein grows, there is a real urgency in finding a sustainable, scalable alternative that doesn't cost the Earth. Insect protein, especially black soldier fly larvae, uses less land, less water to farm, and can feed on a wide range of organic waste to grow. And farming black soldier flies illustrates how we can harness the power of nature's best upcycling machines to revolutionize the way we farm and, therefore, the way we feed the world.

Recording copulation (all in the name of science) of the farmed black soldier flies.

The reputation of the black soldier fly as a superfly has been a long time coming. It was first recorded in its native North America and described by Swedish taxonomist Carl Linnaeus (1707–1778) within the genus *Musca*, which is a type of housefly. For centuries it was considered a compost-dwelling pest to be controlled or eradicated, similar to the species with which it had been originally lumped with. Black soldier flies are not closely related to houseflies, but both have suffered from negative press.

It was the indomitable entomologist Charles Valentine Riley (1843–1895) who began to explore the black soldier fly's life in more detail. This flamboyant artist and illustrator, with a manicured moustache to match, has often been described in North America as one of the fathers of modern entomology. He was a passionate collector of insects, whose knowledge and willingness to share his expertise slowly transformed entomology from a profession of collecting and arranging specimens to a scientific analysis of insect diversity, ecology and the applied management of insect pests. In 1875, Riley stated that 'Insects play a most important part in the economy of nature and furnish us some valuable products and otherwise do us a great deal of indirect good; yet they are chiefly known by the annoyances they cause and by the great injury they do to our crops and domestic animals. Hence some knowledge of insects and how to study them becomes important, almost necessary to every farmer.'

Riley's upbringing was no less transformative. He was born in London to parents Reverend Charles Edmund Fewtrell Wylde and Mary Cannon out of wedlock, and had a different surname to both parents. He and his younger brother George were given the name Riley to conceal this, at the time, socially unacceptable set up. At the age of three, they were moved from their mother's to an aunt on the outskirts

The fine moustache worn by the indomitable entomologist Charles Valentine Riley.

From the 1858 publication on the natural history of insects by Charles Valentine Riley, written and illustrated by the author at just 15.

of London and then on to a 'labouring class family in the status of 'nurse children'." His childhood, although appearing insecure, proved to be inspirational, since he was given the freedom to play along the rivers banks and explore the local parks, where he nurtured his love of nature. When he turned 13, he was packed off to mainland Europe to continue his formal studies and was soon excelling in both art and natural history. At 15 he had already written a book about insects. These insights into his extraordinary early life were all gleaned from the most wonderful accounts by Edward and Janet Smith (1996) who further wrote of the parents' deaths as 'Wylde died in a debtor's prison at 51, and Mary died of dissipation and disappointment at 58'. Death by disappointment might be a lesson to every parent who complains of it when telling a child off!

In 1860, before his mother's death, Riley had emigrated from Europe to the USA, joining family friend George Edwards and family, who had a farm in rural Kankakee, Illinois, 80 km (50 miles) south of Chicago. The friends had emigrated several years earlier to engage in livestock farming, in response to persuasive advertisements in the London papers, joining that great wave of nineteenth-century Europeans eager to acquire land on the 'grand prairie'. Riley initially toiled on the land, and soon gained an understanding of the many trials and tribulations that beset the local famers, including the ravages of insects. But it wasn't long before he moved to Chicago for a less onerous life, taking a writing job with *The Prairie Farmer*, a rag aimed at helping farmers, those people who fed the nation.

Professor Donald C Weber, himself a research entomologist for the US Department of Agriculture (USDA), a position that Riley himself once held, is co-author of the 2019 book *Charles Valentine Riley: Founder of Modern Entomology* and describes Riley as a problem solver. Riley answered the many questions that the workers sent in to him, such as how to improve crop pollination, how to diversify agriculture and how to battle the crop damage by a host of different insects. Riley's reputation was such that he basically became the state entomologist of Missouri where, among his high-profile research programmes, he studied the hungry hordes of Rocky Mountain locusts *Melanoplus spretus* (Walsh, 1866) that invaded many western states between 1873 and 1877. Interestingly he advocated controlling the locusts by eating them and wrote, 'Whenever the occasion presented, I partook of locusts prepared in different ways, and one day, ate of no other kind of food, and must have consumed, in one form and another, the substance of several thousand half-grown locusts. Commencing the experiments with some misgivings, and fully expecting to have to overcome disagreeable flavour, I was soon most agreeably surprised to find that the insects were quite palatable, in whatever way prepared. The flavour of the raw locust is most strong and disagreeable, but that of the cooked insects is agreeable, and sufficiently mild to be easily neutralized by anything with which they may be mixed, and to admit of easy disguise, according to taste or fancy. But the great point I would make in their favour is that they need no elaborate preparation or seasoning.'

This man had ideas decades ahead of their time. But it wasn't just in Missouri that he was known. Riley became hugely influential across the USA, thanks in part to acclaimed books and reports with titles like *Noxious Beneficial and Other Insects*, which were packed with descriptions of new species, life histories and ways to control them. He found splendour in creatures most people found repugnant and devoted many hours to examining some of the most complex of these life histories, including that of the grape phylloxera *Daktulosphaira vitifoliae* (Fitch, 1855), a type of sap-sucking bug, as well as the Rocky Mountain locust, *Melanoplus spretus* (Walsh, 1866) and the chinch bug *Blissus leucopterus* (Say, 1832). His findings on these insects elevated the study of insect biology to new levels, and he also captured the insects he studied in beautiful and detailed hand-drawn images. These books and reports were a real monument to his labours. And as Weber highlights, there is more than an element of showmanship there, for many of the reports were printed on high-quality paper and sent off to other eminent scientists of the period, including Charles Darwin.

In his first report from 1869, Riley describes how he has talked to many a farmworker, keen to understand their 'tiny but mighty insect foes'. Already aware of the damage some insects inflicted on crops, the farmers also sent him insects for him to rear, to further understand tree received from a distance should be examined from "top to stern," as the sailors say, before it is planted, and all insects, in whatever state they may be, destroyed. There can be do doubt that many of our worst insect foes may be guarded against by these precautions. The Canker-worm, the different Tussock-moths or Vaporer-moths, the Bark-lice of the Apple and of the Pine, and all other scale insects (*Coccidæ*), the Apple-tree Root-louse, etc., are continually being transported from one place to another, either in earth, on scions, or on the roots, branches, and leaves of young trees; and they are all possessed of such limited powers of locomotion, that unless transported in some such manner, they would scarcely spread a dozen miles in a century.

In the Pacific States, fruit-growing is a most profitable business, because they are yet free from many of the fruit insects which so increase our labors here. In the language of our late lamented Walsh, "although in California the Blest, the Chinese immigrants have already erected their joss houses, where they can worship Buddha without fear of interruption, yet no ' Little Turk ' has imprinted the crescent symbol of Mahometanism upon the the Californian plums and the Californian peaches." But how long the Californians will retain this immunity, now that they have such direct communication with infested States, will depend very much on how soon they are warned of their danger. I suggest to our Pacific friends that they had better "take the bull by the horns," and endeavor to retain the vantage ground they now enjoy. I also sincerely hope that the day will soon come when there shall be a sufficient knowledge of this subject throughout the land, to enable the nation to guard against foreign insect plagues; the State against those of other States, and the individual against those of his neighbors.

THE CHINCH BUG—*Micropus leucopterus*, Say.

(Heteroptera, Lygæidæ.)

[Fig. 1.]

Few persons will need to be introduced to this unsavory little scamp, but, lest perchance, an occasional reader may not yet have a clear and correct idea of the meaning of the word Chinch Bug, I represent herewith (Fig. 1) a magnified view of the gentleman. The hair-line at the bottom shows the natural size of the little imp, and his colors are coal-black and snow-white. He belongs to the order of Half-winged Bugs (HETEROPTERA), the same order to which the well known Bed Bug belongs, and he exhales the same loathsome smell as does that bed-pest of the human race. He subsists by sucking, with his sharp-pointed

A beautifully printed and worded 1869 report by Charles Valentine Riley which starts 'Gentlemen: I herewith present my first annual report.'

their life histories. But it wasn't just the growers of fruits and vegetables that were concerned with 'injurious insects', American beekeepers in particular had begun to worry, as by the 1880s the black soldier fly had rapidly spread through southern states and honey farmers began noticing these flies invading their hives. Once more flies were gaining, unjustly, a bad reputation.

The beekeepers had discovered the maggots of black soldier flies in the hives, but Riley, as Weber highlights, reassured them they were not feeding on the bees, but their debris. Although the flies resemble wasps, with their tapered abdomens, they were anything but killers. Instead, they were cleaners, getting in and upcycling the rubbish into food. The black soldier fly larvae were soon elevated to harmless scavengers thriving on many kinds of decomposing organic matter, including algae, carrion, compost heaps, manure, mould, plant refuse, and of course the waste products of beehives. The influential Riley had flown the flag for the black soldier fly, recognizing that their apparent acts of destruction were actually an asset.

That list of assets continued to grow into the twentieth century, as we learnt more about the black soldier fly. Unlike houseflies, the adult black soldier flies have greatly reduced sponging mouthparts. They do not regurgitate food along with digestive enzymes like houseflies, and therefore do not spread diseases. Black soldier flies do not fly around as much as houseflies, as they have less expendable energy due to their limited ability to consume food as adults. They are very easy to catch and relocate when they get inside a house as they do not avoid being picked up. They are sanitary, and they neither bite nor sting. Their only defence seems to be hiding. And as a bonus, the colonization of manure by black soldier fly larvae can reduce populations of common housefly *Musca domestica* (Linnaeus, 1758) by more than 90%.

It was in the 1970s that researchers began to value the ability of these tiny munching machines to convert manure into protein, and quickly, outcompeting many of the more nuisance species. A larva will grow 15,000 times its size from hatching by the time it is ready to pupate, which equates to a human baby growing to the size of a blue whale. Aside from the protein production of the larva's furious activity, it also produces another valuable resource called frass – a granulated and odourless residue that can be used as organic fertilizer directly or through conversion by earthworms. Black soldier fly activity also makes manure more liquidy, and thus less suitable for housefly larvae to thrive in.

Professor Jeff Tomberlin of Texas A&M University, USA, is an entomologist focusing on forensic entomology – his lab is called the FLIES facility, as in Forensic Laboratory for Investigative Entomological Sciences. He also focusses on decomposition ecology and yes, you guessed it, black soldier flies. He has known of the larvae's promise for decades and his enthusiasm for its potential as the ultimate upcycler remains unabated. Tomberlin became smitten by them when he was taken by his advisor Dr Craig Sheppard, a man whom he describes as, quite rightly, the grandfather of the field, to a poultry facility where they were to sample them. He describes opening the entrance door to a dimly lit underground space where he is just able to make out a metre-long collection chamber into which poultry manure was raining down from the chickens above. Turning to Sheppard he asks, 'Do we have to go in there?' to which the response is 'no Jeff, *you* have to go in there'. He descends, and wades through the gloomy maggot-infested environment. But instead of being revulsed by the experience, it leaves him with an image of awe. He notices that in a matter of minutes the maggots are ploughing through that manure. It was, he says, a thing of beauty.

Since that dramatic rite of passage, Tomberlin has been alert to the insect's huge potential, in particular, their use in developing much needed sustainable waste management systems. This need has arisen from the growth in global food demands and the subsequent growth in agricultural production required to meet that demand. In a field trial conducted in Georgia, black soldier fly digestion of pig manure reduced nitrogen by 71%, phosphorous and potassium by 52%, and aluminium, boron, cadmium, calcium, chromium, copper, iron, lead, magnesium, manganese, molybdenum, nickel, sodium, sulphur and zinc by 38–93%. This all means the larvae are able to reduce pollution potential by 50–60% or more. Foul odours produced by decomposing manure were also reduced or eliminated by black soldier fly larval digestion, as the species aerates and dries the manure. And the high protein and lipid content of the larvae meant they'd be a useful additive in animal feed and biodiesel production.

But the feasibility of using the black soldier fly as a tool for breaking down waste, and for producing feed, had only been investigated on the small scale until the start of the twenty-first century. Raising black soldier flies at anything approaching a commercial scale wasn't happening. Crucially, no one knew how to get captive flies to reliably mate and deposit eggs. All that changed in 2002, however, when Tomberlin and his colleagues worked it out. The key, they found, was finding the precise mixture of temperature, humidity and especially lighting to stimulate the flies to breed. And so with large cultures to hand it was only a matter of time before the fly's nutritious larvae could be tailored to requirements. Tomberlin aptly states they are what they eat, and their growth rate significantly differs across different treatments. Larvae for instance reared on spent grain grow twice as fast as those fed apples alone, but those reared on the apple and spent grain mixture produce twice as much insect biomass. And by manipulating their diet, the team are able to harvest larvae with different nutritional components. For example, those fed on apples get really fat, that is 60% of their body weight is comprised of fat! If you fed them bananas the fat shifts to protein.

Farming black soldier flies as feed is evolving into big business. The aim is also to exploit the larvae's phenomenal ability to break down food waste in landfills. In a House of Lords report published back in 2018, the amount of food waste produced by the UK was stated to be 9.5 million tonnes. To use one of my favourite measurements of scale, the blue whale, that figure is the equivalent to roughly the weight of 44,000 mature whales, which can weigh up to 219 tonnes each. So not only do these maggots prevent the dumping of all these 'whales' into landfills, they also turn it into an environmentally friendly and sustainable source of protein and fat for animal feed, and possibly more.

At Entocycle, the laboratories contain rearing containers held at precise temperatures and humidity, where Whittaker and colleagues are turning what is naturally a varied and ad-hoc diet into a highly controlled

Digested oranges (above) reduced to grains (opposite) – useful waste for soil fertilizer from the stomachs of the black soldier fly larvae.

process to optimize the production of larvae. It houses, as Whittaker states, the full circle of life, from egg to adult and then back again. For example, each one of the containers has animals of exactly the same age feeding on an exact diet. It's an exact science. They have also developed a machine vision hardware and software package to help others improve

their farms for larger scale manufacturing. And there are many ongoing trials here: the different densities of larvae crowded together, the many types and depths of waste substrate, moisture levels, even the size of the container are all being scrutinized to ensure a consistent conversion of waste food into edible protein on a large scale. And it's a fast process. On day five the larvae are separated from their frass – insect poo – and this too can be used (it is being used by one of the authors on homegrown tomatoes and they are doing great!). You get a real sense of the speed at which these voracious protein powerhouses, given the right conditions, are capable of growing. By day 9, the larvae are 1 cm (nearly ½ in), and ballooning in growth. They are then processed into a protein product within two weeks of hatching. The aim is to produce the UK's first industrial insect facility, Entofarm 1, which will evolve to become a semi-automated production site churning out 2.2 tonnes of sustainable insect protein – and a viable alternative protein source to the increasingly high costs of commercial feeds for livestock and pet food. There are many insect pet food companies today producing nutritionally balanced and environmentally friendly dog and cat food (and some have the best names such as Lovebug Pet food or Bugbakes to name but two). Using larvae as protein would also reduce the currently unsustainable over-consumption of fish. Approximately 20% of the total wild catch worldwide is processed into fish meal and fish oil for aquaculture, so fed back to farmed fish. Aquaculture production has more than doubled in the past 15 years, making the need for sustainable ways to feed farmed fish acute. A 2008 publication by Sena de Silva and Giovanni Turchini state that '13.5% of the total 39 million tonnes of wild caught fish is used for purposes other than human food consumption'. That's around 13,000 blue whales!

In the future, insect farmers envisage creating protein not just for animal food, but human food, too. You could, in years to come, be farming your own protein packets in your kitchen, and to this end a prototype machine from Austrian designer Katharina Unger has been developed, known as Farm 432. It's a multi-chambered box that sits on your kitchen worktop in which you can produce edible fly larvae from your own leftovers.

Unger is CEO and founder of LIVIN farms, a company based both in Vienna and Hong Kong. She undertook the development of these home recycling facilities while studying Industrial Design at the University of Applied Arts in Vienna. Although still in its development, the prototypes show promising results. One gramme (0.04 oz) of black soldier fly larvae can develop into 2.4 g (0.08 oz) of larval protein. That's an amazing conversion. The larvae don't need water, which is great in arid regions,

Farm 432 – making your own protein at home.

and they don't excrete environmentally unfriendly methane (the UN states that 32% of human-caused methane emissions comes from our livestock and this is expected to increase to 70% by 2050).

The process of harvesting protein from one of Unger's farms couldn't be easier. The larvae eat food scraps for about one week until they become fat and juicy, and want to pupate. They then crawl up a ramp and fall down into a harvesting receptacle. Most you can then pick out to eat, but a few are allowed to pupate to turn into adults which then get about their business of producing more offspring. They are basically harvesting themselves. The taste of the larvae is, according to Unger, distinctive, 'When you cook them, they smell a bit like cooked potatoes. The consistency is a bit harder on the outside and like soft meat on the inside. The taste is nutty and a bit meaty.'

The commercial production of soldier fly larvae for human consumption has some hurdles to overcome. The FAO published a report in December 2022 citing that although the regulatory frameworks governing food and food chains had massively expanded over the past couple of decades, there was still a huge absence of legislation or guidance. And even when more than two billion people consume more than 1,900 species of insect in their diet, there is a still a problem of food neophobia, where many folks, particularly in the western world, are reluctant to try new foods, especially if edible larvae are consumed whole.

Is it possible to encourage people to become more adventurous eaters? Research carried out by Kaitlyn White and colleagues at the University of Colorado, USA, has addressed this by looking at how a willingness to eat insects might vary according to individual differences in three areas: a reluctance to eat novel foods (food neophobia); disgust sensitivity (as some people are more disgusted than others); and also current state of hunger. In the first study, students completed online assessments of each of these three variables. They then indicated how willing they would be to eat insects, in the form of roasted crickets, fried worms and insect-based protein bars. Researchers found that the higher scores on the food neophobia and disgust sensitivity scales – but not the hunger measure – were linked to a reduced willingness to consume these items. But would this translate into being less likely to actually eat them? To investigate this, a new group of participants completed those same scales in the lab, and were told that roasted crickets are not only safe to eat but that some people readily consume them. They were then presented with a roasted cricket. The researchers recorded who ate it, and who didn't, and their analysis showed that only food neophobia (not disgust sensitivity or hunger) was linked to actual consumption. In fact, food neophobia scores were a pretty good predictor of who would eat the cricket. So, looking at data across the two studies, while people who scored lower on disgust sensitivity said they would be more willing to eat insects, when it came to it, they weren't. And in both studies, variations in hunger made no difference.

Food neophobia could therefore be the most important hurdle in encouraging western populations to consider insects as food. This would mean finding ways to give people positive experiences of actually eating them, and it may inform future marketing strategies that aim to encourage people to view insect protein as a viable source of nutrition. Realistically it probably won't be the larvae themselves but their protein landing on our plates – cricket flour is already available and is used to make pasta, bread, biscuits and other snacks. Crucially, the case needs to be made to consumers that eating insects is not only good for our health, but also for the planet. Meanwhile, as black soldier fly farms spring up worldwide, these ultimate upcyclers are poised to join honeybees and silkworms as some of the most widely domesticated insects in agriculture. Get ready.

CHAPTER EIGHT

Namib fog harvester

'It may still be argued that to know one kind of beetle is to know them all, or at least enough to get by. But a species is not like a molecule in a cloud of molecules – it is a unique population of organisms, the terminus of a lineage that split off from the most closely related species thousands or even millions of years ago.'

E O Wilson (1985)

YOU WILL BE familiar by now with the sheer number and diversity of insects. We know they get everywhere, even into space, and we know they have exploited many a hostile environment: the deepest caves and at the bottom of lakes, and even the extremes of Antarctica. We now also know that some can tolerate the most arid regions of the world, those defined as having less than 250 mm (10 in) of rain a year. That's quite different to what we experience in London, where on average 615 mm (24 in) falls on every year. That's nothing compared to the wettest part in the UK, where a small village in the heart of Snowdonia in Wales called Capel Curig receives an average rainfall of 2,612.18 mm (84 in). And that's nothing in comparison to the wettest place on Earth, which is Mawsynram in India. According to the *Guinness Book of World Records*, this is soaked in 11,872 mm (467½ in) of rain each year.

Going the other way down the scale to the driest regions, you first have to consider the poles. Oddly, Antarctica, although formed of water,

The Atacama Desert, Chile – the driest non-polar region on Earth.

has an area called the Dry Valleys where no rain has fallen in nearly two million years. When we look at the non-polar regions it's driest in the Atacama Desert in Chile. The town of Arica within the desert is reported to have an annual rainfall of 0.761 mm (0.03 in) per year, while other parts of the desert have not received rain in nearly half a millennium. The last time it rained in the Atacama, Henry VIII was King of England. From South America with Chile we move to northern Africa – Libya, Egypt, Sudan, and Algeria all have very dry regions. Drop down that continent we get to Namibia. Pelican Point in Namibia, famous for its surfing and bird life, is also one of the driest parts in the world despite being within the tropics, with just 13.2 mm (½ in) of rain per year. And you guessed it, insects live in all these dry places. Thanks to the impact of human-induced climate change we are seeing these dry regions expand, and we have turned to study the creatures that live in these extreme environments to see how they manage, and how some seem to survive on water they conjure 'out of thin air'.

Beetles are fantastically adapted to surviving dry places. The order Coleoptera, has more described species any other animal, more than 360,000 and counting. Folks can't get enough of them. Darwin was fascinated by them from a young age and he writes in his autobiography that while out collecting one day he ran out of container pots, but was so desperate to collect a beetle he had found he 'popped' it in his mouth, with an uncomfortable result. 'One day, on tearing off some old bark, I saw two rare beetles, and seized one in each hand; then I saw a third and new kind, which I could not bear to lose, so that I popped the one which I held in my right hand into my mouth. Alas! It ejected some intensely acrid fluid, which burnt my tongue so that I was forced to spit the beetle out, which was lost, as was the third one.'

Darwin was not alone. His co-discoverer of evolution by natural selection, Alfred Russel Wallace, alongside Henry Walter Bates (1825–1892), were stomping through the Amazon in search of them. Francis Polkinghorne Pascoe (1813–1893) may not be such a household name, but he was a bigwig amongst the entomologists collecting across South America, Europe and Africa. And explorer David Livingstone had a relatively unknown liking for them.

Charles Darwin's beetle box.

Livingstone (1813–1873) was born in Blantyre in Scotland but died just 60 years later in Chief Chitambo's village in the then Kingdom of Kazembe, Zambia. He was a legendary figure about whom most of the British public can quote one infamous and hotly disputed line of greeting by American-Welsh explorer Henry Morton Stanley, after years thought missing, 'Dr Livingstone, I presume?' This country boy had grown to become a famous missionary and explorer, a scientific inquisitor and a human-rights crusader. He was a national hero in Britain after his missionary expedition in the 1850s in which he become the first European to cross the width of southern Africa. He vastly improved the knowledge of the continent back home, was the first westerner to describe the Victoria Falls waterfall, and returned with examples of species from some of the harshest environments, never before seen in the UK. One of his boxes of beetles has only recently been rediscovered in the Natural History Museum, 150 years later. It contains specimens from a government-backed expedition between 1858 and 1864, to open up the Zambezi River for commerce. Before you think the Museum is being remiss about its specimens – they were in fact donated to the Museum by a private collector, Edward Young Western,

The beetles of Livingstone, we presume (correctly).

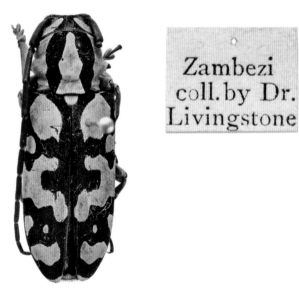

Tragocephala variegata Bertoloni, 1849 – the variegated golden longhorn beetle.

who had purchased these from another individual on the expedition. In the days before digital records, just a line recording it was written in the archives. It's only through a recent and huge digitization programme that its details were revealed.

Max Barclay is Senior Curator at the Natural History Museum, responsible for the team who curate and research the beetle and bug collection, and who also oversees the digitization project for the beetles. He can barely contain his excitement about beetles at the best of times, and the discovery of the Livingstone box was almost too exciting for him! There were 12 different species among the 18 individuals, including *Termophilum alternatum* (Bates, 1878), the aptly named giant predatory ground beetle and several species from the longhorn family Cerambycidae, including the flightless spiny longhorn beetle *Phantasis avernica* (Thomson, 1865) and a darkling beetle or Tenebrionidae – more about that one later.

Livingstone had noted that beetles could be found in even the most inhospitable of environments. And one of the harshest for any beetle, or insect for that matter, has to be the Namib Desert. Stretching along the southwest African coast from Angola through Namibia and on to Cape Town, the Namib is but a few hundred kilometres wide, where giant planes of gravel separate three seas of mobile sand blown in by the Atlantic Ocean. As deserts go, it is considered one of Earth's oldest and gives rise to some of the planet's tallest dunes – up to 250 m (820 ft) high, sculpted by harsh winds. Summer temperatures here reach 45°C (113°F) and night-time temperatures can be as low as below freezing. And perhaps the most extreme aspect of all is that just 15 mm (½ in) of rain falls there each year, sometimes none at all.

Duncan Mitchell is an Emeritus Professor at the University of Witwatersrand in Johannesburg, South Africa, and has for decades been making detailed studies of the Namib fauna. His specialism is in physiology, and he has been looking at how animals cope in such an extreme environment, having recently co-authored an article on the *Fog and Fauna of the Namib Desert: Past and Future* in 2020. To a casual observer, the desert seems devoid of life. That's because most of it lives 200 mm (8 in) within the sand, sheltered from the surface heat and wind.

In fact, fauna only come to the surface when they really have to. But what the apparently barren Namib lacks in rainfall it makes up for with a spectacular inland fog, which appears irregularly on the dunes, close to midnight. This extraordinary event arises from the north and west of Namibia, initially blowing in over the country as a high cirrus cloud drifting over the cold Benguela current of the Atlantic Ocean. The cloud becomes a moist fog in the high atmosphere, and upon intersecting land at a height of about 500 m (1,640 ft), it deposits fog droplets that can penetrate inland as far as 60 km (37 miles). There are other coastal foggy deserts, for example the Atacama, Baja California, Omani and Yemeni. However, no other coastal foggy desert has a combination of the cool coastal climate, large sand dune mass, and gently rising land from the coast, that facilitate this pattern of unique fogging events. It is a life saver for many life forms.

It was South African arachnologist/entomologist Dr Reginald Frederick Lawrence (1897–1987) back in 1959 that wrote of this phenomena in his publication *The Sand-Dune Fauna of the Namib Desert*, with the words, 'with the sinking of the sun and the coming of the cold dank mist which blows in from the sea and deposits a thin mantle of moisture on the sands, a new and different set of life forms arises from the unending sand to pass the night feeding and preying upon each other'.

Lawrence travelled widely for his research; his first collecting trip was in 1923 to Mozambique where his only companion along the wild coastline was his donkey. Later, whilst director of the Natal Museum in Pietermaritzburg, he developed a fascination about the cryptic fauna of the forests of southern Africa and, as was his keen interest all things, he collected many of the cryptic fauna from other habitats such as the Namib. It was Lawrence who first argued for a permanent research station in the Namib to study the life that lived there. With such eloquence he writes about these dune systems seeming as bare as one's hands where humans are '… an unimportant and indeed unnecessary intruder, an utterly unnoticed and insignificant part of the landscape'. Indeed. But he goes on to state that appearances are deceiving, for more than 40 species previously unknown to science including insects,

The Namibian sand dunes (with yours truly posing on them).

arachnids and reptiles had only recently been described from the region. Lawrence finishes his paper exclaiming 'motion pictures of desert life could be made which would far excel Walt Disney's 1953 documentary *Living Desert* in beauty and scientific interest'.

Although the Namib's fogging event is highly variable and unpredictable, and only happens about once a month, the animals know to emerge just before or just as it arrives. It's unclear as to how they accurately anticipate the fog, but Mitchell hypothesizes it may be down to them listening to the wind whilst they are protected down in the sand, since the wind changes direction just before the fog arises. But what can the fog do for them? In 1976, the answer was splashed across the covers of various scientific publications when Namib ecologist Dr Mary K Seely unravelled some of the intimate relationships between the fog and its fauna.

Born in America in 1939, Seely arrived in Namibia in 1967 as a post-doctoral student of Dr Charles Koch, the then Director of the Desert Ecological Research Unit (which became known as Gobabeb). The longed-for research station Lawrence had campaigned for had been established back in 1962 and just three years after Seely's arrival she

The humbug toktokkie *Onymacris marginipennis* – an endemic
Namibian tenebrionid.

took over as Director – a major achievement at a time still plagued by
horrific levels of sexism. The unit is now recognized globally as a centre
of excellence for its research and training. Six years after taking over
as Director, Seely co-authored two papers with William Hamilton 3rd,
one of which was published in *Science* magazine and one in *Nature*.
It was these papers that first described a beetle's extraordinary ability
to harvest and collect water from fog. Welcome the family called the
Tenebrionidae which includes the group locally called the toktokkies.
This is not a taxonomic grouping as such but rather is due to the fact that
they tap their abdomens to communicate with the opposite sex. There
are more than 200 species of toktokkies, of which around 20 have been
supremely adapted to the extreme desert environment and its occasional
nocturnal fogging events.

Seely and her team were the first to observe that some of the beetles
were accessing fog water directly, by wallowing in it. She described how
the beetles would emerge from the dune sand before the fog arrived
– typically a few hours before midnight. They would then clamber

clumsily (because of the cold) up the slip face to the crest of the dunes – that being the part of the dune where fog deposits best –and take up a head-stand posture with the elytra or wing case facing the wind bringing in the fog. And so they would wait, for three to four hours. It was the posture that helped them access the water, allowing fog droplets to deposit on their dorsal carapaces and then drip down towards the mouth, so long as the wind was not too strong and the fog was dense enough. It was an extraordinary discovery. Imagine watching the beautiful activity of these obsessive fog harvesters by torchlight. For 66 nights the researchers did just that.

The beetles appear to be all-consumed when a suitable fog blows in. Their sole aim is to harvest these droplets of water suspended in the air. Despite the abundance of plant detritus and other food above ground, they were never seen to feed during a fogging event, nor did they react to human observers nearby. Because they are cold and lethargic during their fog-basking, they would be vulnerable to nocturnal predators if such predators were prevalent. It is therefore likely that the evolution of their fog-basking behaviour, and indeed the behaviour of other active fog-harvesters, may depend on a low risk of predation, and especially of nocturnal predation.

Lawrence, in his 1959 account of the Namib, writes about physical adaptations of beetles to cope with life on the sand. Some beetles he notes are flattened with short legs to slip into the sand while others are more rotund with long legs for fast running across the hot surface. Indeed, they do look slightly unusual, even Barclay with his beetle devotion describes them as 'funny looking insects'. Hamilton and Seely write in their seminal *Nature* paper about another genus of tenebrionid, *Lepidochora* (Gebien, 1938), the members of which build ridges in the sand to catch water, manipulating the landscape to capture a drink! But it is *Onymacris unguicularis* (Haag, 1875), the head-stander beetle – looking more like a grasshopper than your average beetle with its very long back legs – that has adapted itself for true harvesting of atmospheric fog. Funny looking maybe, but it is perfectly suited for doing these handstands in the fog, and what's more has specially adapted grooves in the exoskeleton that trap sufficient fog moisture for the water droplets to run down. These beetles are fast (the opposite of most tenebrionids), and personal experience of chasing after them but failing miserably is testament to this fact.

The fog-harvesting efficiency of the *Onymacris* Allard, 1885 beetle is down not only to the grooved nature of its exoskeleton, guiding the droplets of pure water to the insect's mouth, but also the exoskeleton's specific chemical make-up, according to Mitchell. Their back is hydrophobic, so it repels water, further encouraging it to run down the insect's back. Although the beetles have never been observed drinking the water, it is known that water is getting into their mouths, as evidenced through the analysis of the chemical signature of the water and insect body fluids, as well as experiments using dead specimens in specially made fog chambers.

Seely and her collaborators scooped up these little marvels both pre and post fog in order to weigh them, thereby determining how much water they were taking in – the highest uptake they found generated a 30% change in body weight in one fog event. Thirty per cent! It's an extraordinary amount of fluid, hugely efficient for an animal as small as a penny to sustain until whenever the next desert fogging event takes place. It is the equivalent of us drinking 20 litres of water (or just shy of

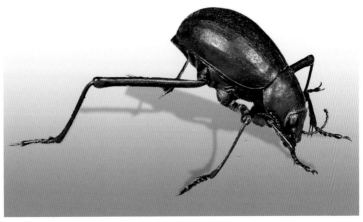

Onymacris unguicularis – the head-standing, fog-harvesting tenebrionid beetle.

27 bottles of wine) in one go. This rapid acquisition potentially creates problems for storage and regulation of body fluids, but beetles have solved this in enterprising ways. Research shows that they keep this large amount of water separate from the rest of their body fluids. The fog water is incredibly pure, with virtually no electrolytes, so it has to be kept away from the rest of their circulatory system so as not to chronically dilute the animal. In the interval between fogs, as the insects become drier and drier, the water is then gradually introduced. Another solution is to isolate the fog water internally and then gradually to add osmolytes to it, only then allowing it to mix with other body fluids. This storage strategy has been reported for *Onymacris unguicularis*, which retains its harvested fog water in its gut. The effectiveness of this process is borne out in long-term studies of the population density of beetles in the Namib – fog-harvesting beetles maintain their numbers during dry periods, compared to others that lack this adaptation.

Other potential water-collecting surfaces, from other species, have also been investigated. One in particular, the bumpy back of a *Stenocara* Solier, 1835 beetle, piqued the interest of bioengineer Dr Andrew Parker (see chapter six), and Chris Lawrence of Mechanical Services Sector, QinetiQ, a defence technology company, who published their findings in *Nature*, in 2001. Parker describes the back of a *Stenocara* as a mountain range with many peaks and valleys. The bumps or tops of the mountains are smooth with no waxy material on them, while the valleys and the sides of the mountains are lined with it. The smooth tops attract water and are super-hydrophilic (water loving) while the sides and bottoms are super-hydrophobic (water fearing). That bumpy surface, along with the wax, causes the accumulating water from the fog to ball up, that is, as the water hits the beetle's back, it is propelled from those valleys to the peaks where a droplet forms, one that's large enough and heavy enough to roll down to the beetle's mouth. It is the combination of both a hydrophilic and hydrophobic surface that causes the drop to form.

A later publication identified their model species as *Physasterna cribripes* (Haag-Rutenberg, 1875), another tenebrionid, which looks very similar to *Stenocara* with its bumpy back, and as such their investigation into the impact of bumps is still pertinent. This paper

The fog-harvester *Onymacris bicolor*. The important feature of the beetle's carapace for fog-harvesting is not the colour (unusually white here) or shape of the elytra, but whether the carapace is grooved.

by Thomas Norgaard and Marie Dacke, of the University of Lund, Sweden, published in 2010, looked at the water-collecting efficiency of the smooth-backed *Onymacris unguicularis* and another known fog-harvester *Onymacris bicolor*, along with the bumpy *Stenocara gracilipes* Solier, 1835 and *Physasterna cribripes*. Only the first two have been determined to harvest in nature but all offer the potential for bio-inspiration.

So what form could that bio-inspiration take then? *Star Wars* fans will remember Luke Skywalker's moisture farm on the desert planet Tatooine – a series of huge white towers called moisture vaporators, which were used to capture moisture from the air. But could they work in real life, and help tackle growing global concerns of water scarcity? Armed with knowledge about the unique properties of the *Stenocara* beetle's geometry, Parker began devising fog-capturing devices in sheet form, using 3D printing techniques. This technique prints layer upon layer to construct a three-dimensional object, hence the name. Parker has been looking at using hydrophobic inks on a hydrophilic surface to recreate the beetle's strange but useful morphology. And you don't have to limit this to the size of the beetle. They have hung up sheets in the Atacama Desert, the most arid of deserts but which can also be subject to fog, to harvest water from. The water drips down the sheet in to containers below. A single harvesting sheet averaging 50 sq m (538 sq ft) can, with optimal conditions, gather 1,000 l (220 gallons) of water in a single fogging session.

These are impressive figures, but the harvesting efficiency of these designs becomes compromised with an increase in sheet size, as the fog droplets have further to travel before reaching the collecting chamber,

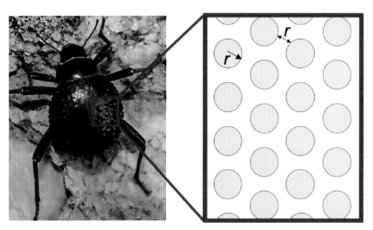

The bumpy back of the *Stenocara* beetle.

and so are more at risk of evaporation, particularly in warm windy conditions. Assistant Professor of Mechanical Engineering Dr Kyoo-Chul Kenneth Park of Northwestern University, USA, has been looking at ways of overcoming fog-harvesting limitations at an industrial scale. He's a self-professed *Star Wars* fan – on his desk sits a Lego model of one of Luke Skywalker's moisture vaporators – inspiration perhaps to push this futuristic technology still further! Park has been adopting a multi-bioinspired approach to the problem, drawing on clever strategies employed by particular flora and fauna in both desert and rainforest environments. He's developed fog-harvesting designs that use a surface structure with slippery asymmetric bumps. These not only incorporate the Namib Desert beetle's efficient formation of droplets, they also borrow structures seen on the asymmetric spines of cactus plants, and the unique slipperiness of the rimmed surface of a pitcher plant. When combined, these features make for a fast continuous, directional transport of water in a minimal amount of time, thereby reducing water loss through evaporation. This strategy could, Park hopes, enable a wide range of water-harvesting applications on a larger scale.

Park also wants to apply his work to a new frontier – smog harvesting. With 4.2 million people dying each year from outdoor air pollution, according to the World Health Organization, extracting smog from the air in major cities could help tackle growing health and environmental challenges. Smog is a combination of fog and smoke, and capturing it is currently a far more difficult process than extracting water from fog, because water in the air clogs up the smog-filtering devices. But Park argues that instead of condensing fog, it is feasible to capture smog in the same way. Not just that, he is also exploring how to use smog-harvesting technology to tackle another one of the great challenges that has eluded many technological efforts – reducing the discharge of brine into the ocean by desalination plants. He's hoping the technology will soon prove to be more than mere science fiction.

But back in the present, fog-capturing devices inspired by a beetle's back are becoming a major industrial enterprise for sourcing freshwater in drought-stricken areas of our planet, where the right mix of geography and climate can be found. But with global warming, the fogs are disappearing. And with that, much of the Namib fauna is threatened with extinction. Despite this climate uncertainty, beetles have another trick up their sleeve that could point to their desert survival in the longer term, were the chances to harvest fog to become even less frequent. It lies in the design of the beetle's bottom. The beetle's rectum does the same job as the rectum of most mammals or insects – absorbing nutrients and water from bodily waste before it is expelled. But beetles do it better than other species, and as a result, beetle faeces are practically bone-dry. A big part of how they do this is down to the design of the beetle's organs. Unlike in mammals, the beetle has kidney-like organs called malpighian tubules (named after Malpighi who first drew them) closely applied to its rectum, and the entire structure is encased in a chamber by a membrane. This allows the animal to generate high salt concentrations in the tubules, enabling the beetle to extract all the water from its faeces by osmosis and recycle the moisture back into its body. It is also able to open its rectum in high humidity, allowing it to take water in and, again, almost fully absorb it into itself.

While scientists have been aware of this unique approach to water consumption for more than a century, only recently has Dr Muhammad Naseem and a group of collaborators from the University of Copenhagen, Denmark, as well as the University of Edinburgh, UK, and the University of Glasgow, UK, been able to shed light on the underlying mechanisms involved. They studied red flour beetles *Tribolium castaneum* (Herbst, 1797), considered model organisms because they are easy to work with and share biological similarities with other beetle species. The researchers discovered a gene expressed at 60 times higher levels in the beetle's rectum than in other parts of its body. This gene led them to discover a distinct group of cells positioned like windows between a beetle's malpighian tubules and its circulatory system or blood, which have been found to play a crucial role in the beetle's water absorption process through its rear end. As the beetle's tubules surround its hindgut, these cells pump salts into the kidneys, allowing them to take in water from moist air through the rectum and into the beetle's body.

With climate change, it is conceivable that opportunities to tap into the ethereal power of fog may slip beyond our reach. But the legacy of both the beetle bump geometry, and the grooved back of the head-stander *Onymacris* beetle, will live on. Labs have been developing a host of applications that mix water-attracting and water-repelling surfaces, such as windows and mirrors that don't fog up, or self-filling water bottles. An intriguingly simple model for a self-filling water bottle, created by South Korean industrial and graphic designer Kitae Pak, consists of a stainless-steel dome that adopts the simple groove principle of the *Onymacris* body. The Dew Bank Bottle is meant to be placed outside in the evening. In the morning, when the surrounding air begins to warm, water droplets condense to the steel surface of the bottle, which has become cold overnight. The dewdrops collected are then channelled down grooves in the surface design, to an enclosed circular holding-chamber. The designer expects that this device would gather enough moisture for a full glass of water per use.

The beautifully designed self-filling water bottle next to the beetle that inspired it.

Parker highlights how the beetle bumps can help recapture valuable water from air-conditioning units. In Japan for example, large air conditioning systems within city buildings consist of a central tower, resembling and acting like an exhaust that sends water vapour out into the atmosphere. This causes the urban air to rise in temperature by up to two degrees, which is both harmful and wasteful. If, as he suggests, we instead coat the towers with the 'beetle bumps' we could recapture the water, but also prevent the heat entering the environment that would significantly help reduce the temperature of the urban environment. Let's drink to that!

CHAPTER NINE

Bee brain intelligence

We dance. We prance. We waggle. We wiggle.
Come glance at our stance while we jiggle and wriggle.
Our dance round and round, shows where flowers are found.
And our figure eights, show where pollen awaits.

Douglas Florian (2012)

Bees are thought of as rather intelligent little creatures. Many of us can remember being taught that they can communicate complex geographical information to each other with a wiggle of their behinds. But it's not just their ability to communicate that is fascinating us, it's also their ability to learn. And in learning about them, we are also learning about other insects, and intriguingly, more about how to navigate our own world.

Humans have conflicting attitudes towards bees. Some of us love these fluffy flying packages and understand their value as pollinators, while others have a morbid fear of them. Writer Jeff Lockwood, author of *The Infested Mind: Why Humans Fear, Loathe and Love Insects* has explored these polarized views, and how they have come about. He recounts an article by Dr Dewey Caron, Emeritus Professor of Entomology and Wildlife Ecology at the University of Delaware, USA, and all-round lover of bees. Caron is active in bee education and publishes often on how to get the public to love bees and other insects,

The waggle dance of the honeybee.

but has also published a story about a man in Africa who was attacked by rather a lot of them, 'The unfortunate man jumped into the shallow river as the bees literally coated his body… he began to sicken from the effects of the venom. Vomiting, he managed to move into deeper water… his head ached badly. He suffered from diarrhoea so intense that he was incontinent.'

Sometimes scientists help reinforce rather than reduce prejudice against some animals and all that good will to bees was undone in a few short sentences. It is true, however, that bees can inflict pain, and in the worst-case scenario, death. The Centers for Disease Control reported that, in the USA between 2000 and 2017 there were a total of 1,109 deaths (a yearly average of 62) from hornet, wasp, and bee stings.

Bees, wasps and hornets belong to the Order Hymenoptera (from the Greek meaning 'membranous wings'), a group packed full of stinging, venomous insects as well as some more docile ones. The stinging species are found within the infraorder Aculeata (part of the suborder Apocrita) – a diverse and species-rich group, with more than 70,000 species described to date. The name refers to the morphological modification of their egg-laying tube, their ovipositor, into a sting. However, this is definitely not the majority of cases in this group, as not all sting. Some either keep the ovipositor for its primary function or lose it altogether. Because the sting is a modified ovipositor it means only the females are stingers. And the majority of these are quite happy in both stinging and laying eggs, stinging, and laying eggs, and so on. Many of these are solitary creatures as well, living, working and rearing by themselves.

Our knowledge of stings dates back millennia. The first recorded case of a fatal sting, from anaphylaxis, was a King Menes, a fact reported in medical literature. Sadly, this is more than likely untrue, as firstly there was probably no such king and secondly, the person who interpreted the two hieroglyphs telling of this fatal jab probably got it wrong. The interpretation was made by Lieutenant Colonel Laurence Waddell (1854–1938). Hailing from Scotland, he became an Indian army surgeon, as well as both a Professor of Tibetan Studies and Professor of Chemistry and Pathology, all of which sound laudable on paper. But much of what he wrote was heavily biased by his imperialistic attitude or, as in this case, wishful thinking! No other Egyptologist agrees with his wasp sting reading, but like many myths associated with insects and their dastardly behaviour, the story is still being told.

One sting fact we can verify, is the use of stinging insects in human warfare, to defend against an attacking force or drive out the enemy. By 2600 BC Mayans had found out how to set bee-laden booby traps, and the use of traps and bee-cannons has been seen across the globe – both on land and on the seas. Indeed hives have been recorded being thrown from one ship to another! Lockwood writes about this early kind of biological warfare, where around 100,000 years ago our ancestors were happy lobbing bee, wasp and hornet nests at each other. In the sixteenth century, French essayist and philosopher Michel Eyquem, Sieur de Montaigne, known as the Lord of Montaigne (1533–1592) wrote a chapter in his *Essays: Book II* about the 1513 siege of the City

The slinging of beehives by the besieged citizens of Tamly at the invading Portuguese army in 1513.

Apis mellifera – the honey-bearing bee. A 'domesticated' goddess.

of Tamly, in the territory of Xiatine by the Portuguese army. It was not going well for the inhabitants, until they devised a cunning plan involving the throwing of bee hives (of which they had many) over the city walls. This, combined with fires, 'drove the bees so furiously upon the enemy, that they gave over the enterprise, and truss'd up their baggage, being not able to stand their attacks, and endure their stings'. Not a single man was lost thanks to these angry females.

The anger of a hive is widely known from the honeybee, *Apis mellifera* (Linnaeus, 1758), which belongs to the eusocial Hymenoptera. Eusociality (meaning good and social) offers many benefits to an 'organized' society, through co-operation and division of labour. Here there is one queen (to rule them all), who is the only egg-laying female in her hive. Her many offspring include workers who can sting, and thus defend, but are not capable in most circumstances of laying eggs. Unlike the eusocial wasps, that will actively go out and hunt prey, using their sting to subdue a meal, honeybees use their stingers for defence, acting for the 'greater good' even if some die in the process. The sting of a bee is designed for other insects, it is a barbed harpoon attached to a venom sac. But mammals, including us, have a thick skin within which the sting gets wedged. As she pulls away, her body is ripped apart leaving the still pumping venom sac behind – her death a sacrifice for the greater good of the hive.

Honeybees are also rather clever. We thought ourselves very intelligent for using the very nature of these creatures to our advantage. Any intelligence that humans might have had in adopting bees for warfare certainly wasn't thought to apply to the minds of the bees themselves. They were thought to be governed by instinct – not insight. Recently we have patted ourselves on the back for creating 'sniffer bees'. Sniffer bees (and wasps) have been trained, yes trained, to seek out different smells, aka chemicals such as narcotics or explosives, in the same way that we train other animals such as dogs. By a process called classical conditioning – the rewarding of a response to an odour – researchers have been able to train bees in a remarkably quick time, just five minutes (it takes up to 8 months to train a dog to do the same thing). And 10 years ago, for another BBC Radio series I went to a lab in the UK to actually see this for myself. Inscentinel was a firm based at Rothamsted Experimental Centre (the same one founded by Margaret Fountaine's Uncle) that, although no longer operating, was then researching the pavlovian responses of bees to chemicals. When a sugar solution with the chosen chemical was placed in a fume head with the bees, they would stick their tongues out. Just like Pavlov's dogs who produced digestive juices when the bell was rung, the bees would, after this initial training, stick their tongues out when they smelt the chemical again, having associated with a sugar reward. Sniffer bees have been used in all sorts of environments including war zones where they are used to detect land mines.

Pioneering biologist Charles Henry Turner.

For centuries the idea of intelligent insects was considered to be a contradiction in terms. Most of the historic literature about them describes them as having a spontaneous reaction to life, governed by instinct. Neuroscientist Professor Martin Giurfa at the University Paul Sabatier of Toulouse, France, studies perception and learning in honeybees, and believes that since insects have small brains, any cognitive aspects were previously dismissed. His research has found that, yes honeybees have miniature brains but they can learn new behaviours based on their capacity to make links between olfactory or visual stimuli, revealing their unsuspected cognitive capabilities, which require an explanatory level beyond that of elemental learning. So they are far more complex than 'see flower, fly to flower'. To reduce insects to just instinctive creatures, he and many others argue, is wrong.

Darwin was famously generous in attributing intelligent behaviour and mental abilities to animals, yet his mid-nineteenth century musings were based largely on observation and inference. But a pioneering young African-American biologist (and civil rights activist) Charles Henry Turner (1867–1923) agreed with Darwin's assertion that humans were not the only intelligent animal on the planet.

Turner began a host of investigations in the 1890s that were in sharp contrast to prevailing ideas about animal behaviour and cognition. In his 1911 paper, for example, on the *Experiments on Pattern-Vision of the Honey-Bee*, he states that 'Any lack of response to a stimulus does not mean that a stimulus has not been noticed by the insects; but that, to them, it has not yet acquired a meaning'. Turner would go on to publish more than 70 papers on intelligent problem solving in many invertebrates, in particular bees. Most scientists today would be happy to have had that kind of turnover, but what made it even more amazing was that he wasn't even based at a university for the majority of his career.

Dr Jessica Ware is an African-American evolutionary biologist and entomologist working at the American Museum of Natural History in New York, USA. Ware is not only a great entomologist and researching the complicated story of insect evolution, but she is also active in the fight against sexual and racial prejudice in science. And she knows a lot about Turner. Ware says that the fieldnotes Turner took were incredibly detailed, reflecting his immense patience. It took painstaking research in often adverse conditions for him to tease out individual variations in bee behaviour. This is someone who sat for hours on end, in the baking hot sun, maybe wearing a suit as was the norm then, and describing every single bee he saw, including the position of its legs and the flowers it was visiting.

Turner had been raised in Ohio just after the civil war had ended, a turbulent time but one that did not hold him back. He showed a keen and inquisitive mind early on in life, as well as an avid interest in nature, and was encouraged by his parents Thomas Turner, a church custodian, and Addie Campbell, a nurse, who would support him throughout his school years. Ware muses on what led Turner to have such gumption, considering racial tensions at the time. The infamous Jim Crow laws forcing racial separation in all public facilities, from libraries to schools, had been implemented in the 1870s and continued on until the 1960s, when many within the African-American community went from slavery to segregation. In 1896, Turner married Leontine Troy, and they had three children: Henry (1892–1956), Louise (1892–94?) and Darwin (1894–1983), who was indeed named after the great man himself. With a young family, and a supportive wife (whether there was any wider support it is not known), he obtained a degree from the University of Cincinnati in 1892. This is notable in itself as he was the first African American to graduate from this university. But if that wasn't enough, by the time he graduated Turner already had three publications on a rather eclectic range of subjects: *A Few Characteristics of the Avian Brain* and *Grape Vine Produces Two Sets of Leaves During the Same Season*, both published in *Science* (its first black contributor); and *Psychological Notes upon the Gallery Spider – Illustrations of Intelligent Variations in the Construction of the Web* in the *Journal of Comparative Neurology*.

Individual spiders, Turner observed, seemed to adapt how they built their webs depending on the geometry of the available space and in the prey they were interested in, writing, 'We may safely conclude that an instinctive impulse prompts gallery spiders to weave gallery webs, but the details of the construction are the products of intelligent action'. It was an interpretation of animal behaviour that would underlie his entire body of work, and was a passport into the weekly meetings run by his mentor Professor Clarence Luther Herrick (1858–1904), a researcher focused on both geology and comparative neurology.

Lab meetings are a common process by which all staff get together to discuss results and findings and throw around new ideas and proposals. They can range from the hugely interactive, with senior staff members being incredibly supportive of the junior staff or the exact opposite. The way that Herrick ran his was more like a high tea, with clothed tables laden with sandwiches, and, as was tradition at the time, whites only. But Herrick asked the other members if they minded Turner attending to which they replied no, and so he was invited to discuss '*science and the intellectual pursuit of it*' – as Ware puts it '*he was no longer being recognised by his race or creed*' but rather by what he could offer mentally.

Such a positive start to his academic career soon faded, however. He initially started his PhD at Denison University shortly after graduating in 1893 but this was discontinued in 1894. He lectured for a while at Clark University, the four-years liberal arts college for the African American students (now Clark Atlanta University) but sadly the exact dates of this were not recorded. In 1895 Turner's wife Leontine died, and he was left to raise their two boys, their daughter having also passed away previously. He struggled to find a university position and so became the principal at a College Hill High School in Cleveland, Tennessee in 1906, remarrying soon after, after which he moved institutes again to Haines Normal and Industrial Institute in Augusta, Georgia, the area's first black school. Turner was still researching in his spare time, focusing more and more on insect behaviour, and to this end received his PhD in Zoology from the University of Chicago in 1907 – only the third African American to have done so. You can but imagine the frustration that despite the PhD, and with 20 publications under his belt, he still could not secure an academic appointment. Neuroscientist Giurfa in France has also studied and written about the life and work of Turner, and hints that it's possible his isolation from standard academic institutions promoted his originality.

Turner's original thinking continued to be played out when in 1908 he turned to teaching at Sumner High School, an African-American

Vol. XV.   *November, 1908.*   *No. 6.*

BIOLOGICAL BULLETIN

THE HOMING OF THE BURROWING–BEES (ANTHOPHORIDÆ).

C. H. TURNER.

INTRODUCTION.

The researches about to be described were conducted for the purpose of determining how the burrowing bees compare with the ants and the mud-dauber wasps in their method of finding the way home. During most of the month of August, 1908, from five to ten hours a day were devoted to this study. This made it possible to conduct several series of experiments. Since all of the series led to similar conclusions, only two of them will be recorded. The majority of the experiments were conducted upon a species of *Melissodes* Latrl., many nests of which existed in an abandoned garden of the Haines Normal School.

SERIES A. EXPERIMENTS ON MELISSODES.

These experiments were conducted in a deserted garden. Before beginning the experiments proper, numerous preliminary observations were made for the purpose of obtaining information that would be helpful in conducting and interpreting the experiments.

Bearing in mind Bohn's assertion that the flights of certain Lepidoptera are anemotropisms and phototropisms,[1] much attention was given to the flight of these bees.

When these anthophorids are busy at work, the flight is certainly neither an anemotropism nor a phototropism, for neither the movements nor the orientation of the body bear any constant relation to either the direction of the wind or to the rays of the sun.

[1] M. Bohn, "Observations sur les Papillons du Rivage de la Mer," *Bull. de L'Institut Général Psychologique*, 1907, pp. 285–300.

247

The eloquence and insightful experiments of Turner.

182

school in St Louis where he stayed until ill-health forced him to retire in 1922. It was during this period, without state-of-the-art lab facilities or access to libraries, that he turned to fieldwork at home or school gardens with his pupils, feeding his curiosity for all things bee.

Between 1908 and his death in 1958, Turner published 41 papers on invertebrates. And it appeared he loved bees the most, all of them, not just the honeybees. He wrote many research papers on parasitic bees and solitary bees and was the first to describe the nuptial flight of long-horned bees *Melissodes* Latreille, 1829 in a publication called *The Sun-Dance of Melissodes*, a title as romantic as the act itself. His writing was as engaging as the subject, with him setting the scene of the activity, 'The time was the month of August; the place, an abandoned garden in Augusta Georgia'. It sounded like a teenage triste, not the previously unknown mating behaviour of a hymenopteran. Turner further writes in the voice of the female bee proclaiming, 'My behaviour is much more than a complex anemotropisms and phototropisms, for my homing is controlled by memory pictures of the environment of my nest'. Who knew that bees were so scientifically minded. At the same time, he also showed that solitary bees dig little holes in the ground and that they learn to remember where they are by studying the landmarks around them.

Turner confirmed this through two series of experiments. The first was conducted in a deserted garden where he manipulated the visual clues around the holes dug by the bees using classic scientific apparatus – pieces of paper and watermelon rinds. The second was the same as the first but with a smaller unidentified species of bee. By moving things around he determined that bees could learn their landscape, and that their search behaviour could only be explained if the animal retrieved from a memory store the information that guided its decision. Turner wrote, 'By a process of elimination, the most consistent explanation of this behaviour is that burrowing bees utilize memory in finding the way home and that they carefully examine the neighbourhood of the nest for the purposes of forming pictures of the topographical environment of the burrow'. These insects were operating above and beyond instinct.

Turner was using the scientific method, a series of repeatable experiments to enable him to produce some experimentally sound results which he extended into investigating the ability of honeybees to see colour. The German-Austrian physiologist and Nobel-Prize winner Karl von Frisch (1886–1982), is widely credited with demonstrating colour vision in honeybees, and published a paper on his findings in 1914. But it was Turner who devised detailed experiments investigating this phenomena four years earlier as 'a matter of much theoretical importance for the correct interpretation of the relations of insects to flowers'. Rather than work with honeybees from an apiary, he sought to attract wild ones for a series of experiments he performed over the course of five consecutive summer days.

As he details in his 1910 essay *Experiments on Colour Vision of the Honeybee*, he carried out three experiments in O'Fallon Park in St Louis Missouri. He designed various red, green and blue disks, coloured boxes, and 'cornucopias' into which the bees were trained to fly, and placed them among blossoms of sweetclover, covering some with honey to attract the bees. Unfortunately, unbeknownst to Turner, the card with the honey reward was coloured red – a colour that bees are blind to. Although there is now no question bees can see such stimuli it's probable his experiment was scrutinizing achromatic vision, which allows bees to distinguish between various shades and tints rather than true colour. Despite this, he was still able to show that bees used both visual and olfactory cues in seeking out their nectar rewards, stating that, 'While proving that bees have colour-vision, these experiments throw no light upon the colour preferences of insects. That has not been the purpose of these researches. The aim has been to answer the question, Can bees distinguish colours? The experiments seem to demonstrate that foraging bees have precepts and that two factors which enter into those precepts are colour sensations and olfactory sensations.'

But because he had observed honeybees fly directly to his red vessels when the vessels were in the shade or direct sunlight, Turner reasoned that bees did not rely on discriminating shades of grey but could recognize colours. Furthermore, Turner was also explicit in the way in which he conceived the behaviour of the bees. He explained the choice of his artificial stimuli in terms of meaning acquisition. 'To the bees these things had acquired a meaning; these strange red things had come to mean honey bearers, and those strange green things and strange blue things had come to mean not-honey bearers.' In this way, Turner anticipated fundamental principles of learning through association. Unlike Von Frisch, who was reluctant to make claims about the mental abilities of animals, Turner refused to see bees and other insects as simple reflex machines driven by spontaneous reactions to environmental stimuli. For him, behind every insect's decisions lay learning, memory and individual variability.

This was also borne out with yet another set of experiments. Using the same set up as before, could honeybees be shown to recognize and distinguish between patterns? The difference this time was that the disks had vertical or horizontal stripes, coloured black and white, or red and green. He performed nearly 20 experiments, varying the way he displayed these stripes and showed that bees learned to choose those whose pattern had been previously associated with honey. Confronting vertical versus horizontal striped patterns yielded no doubt – despite showing the same colours (for instance green and red) and the same spatial frequency (the same stripe spacing), bees preferred a previously rewarded vertical-stripe pattern and ignored the horizontal-stripe pattern. Turner concluded that bees learn and recognize colour patterns and that in doing so they also learn the spatial distribution of colours, that a vertical and a horizontal grating of identical colours are not the same. In considering the results of his experiments, he believed that bees may be creating, in his words, 'memory pictures' of the environment. And he was right.

These cognitive perspectives on animal behaviour, unpopular at the time in a scientific environment dominated by behaviourist views, underline just how advanced Turner was in his time. As an interesting aside, Turner admired George John Romanes (1848–1894), an evolutionary biologist whose 1883 book *Animal Intelligence* contains the following, 'Starting from what I know subjectively of the operations of my own individual mind, and the activities which in my own organism they prompt, I proceed by analogy to infer from the observable activities of other organisms what are the mental operations that underlie them'. Turner's respect for Romanes was so deeply felt he named his second son Darwin Romanes Turner.

A bumblebee pulling a string to get to a nectar reward.

Turner's analysis of honeybee foraging and orientation anticipated our current interpretations of insect behaviour, and it is this that modern-day researchers are rediscovering and continuing to build on. In his lab at Queen Mary University in London, UK, Professor Lars Chittka, a German entomologist, is examining what bees are capable of learning if taken out of their natural environment, and if they can even anticipate outcomes of their own actions, in order to understand more about sensory systems and cognition. And he has one of the best lab set ups I've ever seen!

Chittka's Bee Sensory and Behavioural Ecology Lab is testing the ability of bees to gain access to artificial flowers that are concealed under low glass panels. By pulling on strings that are attached to nectar rewards, the bees can drag the reward out from under the glass. Yes, the bumblebees *Bombus terrestris* (Linnaeus, 1758) were learning to operate string. Although it might look amusing, there is a deeply interesting undercurrent running through this research. Because the string can be either long and therefore harder to pull, or short, and thus easier to pull, the bee has to choose which one to go for. As Chittka states, 'They have to look at the strings and make a judgement about which is the best one to pull for their rewards'. His experiments don't stop there, for indeed there is an even more complicated one where some of the flowers are not actually attached to the string – so the bee now has to compare a short but fruitless string with a long but rewarded string. In the latter case it is beneficial to pull the longer, harder, string. And while this is all going on the other bees are observing the work – there is no deliberate effort to pass on information but only as a by-product of watching a 'skilled demonstrator' as Chittka calls them.

This much learning in so small a brain is one of several important insights bees can give engineers, and is fuelling a new generation of bee-sized robots. Assistant Professor Elizabeth Farrell Helbling of Cornell University, USA, heads the RoboBee project, developing fleets of mini flying machines that aim to replicate the bee's intelligence along with their versatile flight abilities. The idea of developing tiny robotic 'creatures' to help humans is a long-told story. The 1966 film *Fantastic Voyage* sees CIA agent Grant (Stephen Boyd) and his team miniaturized to get into the brain of the scientist who, thanks to an escape plan that went wrong, now has a brain clot, and so can't tell them how to perpetuate the miniaturization process – plot spoiler, they succeed. More recently we have had *Innerspace*, *Honey I Shrunk the Kids* and its sequels, and *Ant-Man* (a favourite with entomologists), all riffing on the theme, and there are many reasons why the idea of miniature humans, and more realistically robots, is attractive.

Helbling's RoboBee is the size of a penny and weighs only 90 mg (that's about 10 houseflies). It joins a small battalion of tiny robots that includes one named black hornet. This fits into a human hand and is able to cover 2 km (1¼ miles) in 25-minute bursts. There is

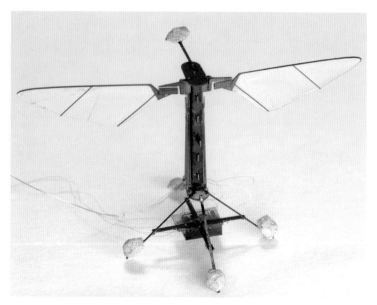

The RoboBee, a mini flying machine that aims to replicate the bee's intelligence along with their versatile flight abilities.

also a RoboFly, which as you may guess is similar to the RoboBee, and powered by a laser beam.

Tiny robots can be extremely useful for tasks where you need to obtain a lot of information quickly from a large area that is too dangerous for humans to enter or cross, such as a collapsed building or enemy territory. Inspecting a gas leak, searching for survivors and navigating hostile terrains would require the robot to have a degree of learning ability, so it could determine all the novel objects it encounters. Helbling has been successfully trialling these RoboBees in the lab and is excited about moving towards field trials. The aim in the early stages will be to embed them with enough intelligence that they can store and process sensory information, that they can discover more about the world around them and easily navigate through an environment. They will have to learn how to make accurate decisions, to discern for instance what objects to seek out or avoid.

As we continue to draw on the bee's extraordinary abilities to learn, could the small brain of a bee also plan, and imagine? Chittka has many thoughts on this and comments, 'We have more and more fragments of evidence that bees don't just live in the present, but they can project into the near future, and that there may be emotions in bees – more and more pieces of evidence point to the conclusion that there is a form of consciousness in bees'.

You can't help but wonder what Turner might have achieved if he'd been given more academic support and resources. The entire field of animal cognition, and tiny brain capabilities, may have developed very differently. Nevertheless, he achieved a great deal, as did Margaret James Strickland Collins (1922–1996) – the 'termite lady', another African-American entomologist and civil rights activist who balanced both during her life time. Both Turner and Collins achieved so much in the face of adversity, designing sound scientific investigations, as Ware phrases it 'a real testament to their work ethic'.

It takes exceptional individuals to extract knowledge from the world of tiny insects. Just as Turner's nature was to work hard and overcome obstacles, so the bees and other insects work hard to navigate the many obstacles and challenges they face. The lesson here perhaps is don't judge a bee by its sting. Now that is something to think about.

CHAPTER TEN

The nerve of a cockroach

Scuttle, scuttle, little roach-
How you run when I approach:
Up above the pantry shelf,
Hastening to secrete yourself

Christopher Morley (1890–1957)

T HE COCKROACH. It likes to eat the same food as us, it likes to live with us and it is especially troublesome in kitchens. In my case, there is at least one under the bath. Cockroaches are supreme colonizers, and we have a funny relationship with them. Some people dislike them or have a phobia against them (katsaridaphobia) due to their perceived association with filth. While others, like me, are intrigued. But few realize there is something rather amazing about these creatures, too.

Cockroaches come from an old lineage, with ancestors that evolved in the Cretaceous Period, back in the time of the dinosaurs. They are grouped with the termites, in the order Blattodea and in fact termites are really just very social cockroaches! There are approximately 4,400 species of them, but most folks will only have encountered three dominant synanthropic species, those that live alongside humans – the

Cockroaches – troublesome in the kitchen.

American cockroach *Periplaneta americana* (Linnaeus, 1758) which is in fact native to Africa and the Middle East, the German cockroach *Blattella germanica* (Linnaeus, 1767), which originated in southeast Asia, and the oriental cockroach *Blattella orientalis* (Linnaeus, 1758), which originates from the Crimean Peninsula – so that's a hat-trick of geographically incorrectly named species! These along with a few others have realized that humans are a good bet for something to eat and somewhere to stay, and have taken to living alongside us whether we want them to or not. The rest of the Blattodea are off doing their own thing in forests, in caves and many are aquatic (they are really good at holding their breath).

We know of two cave-dwelling cockroaches from the Cretaceous. Hemen Sendi and co-authors published a paper in *Gondwana Research* in 2020, looking at these specialized creatures, how they had adapted their bodies, evolving small eyes and a reduced overall body size, in order to live in this highly specialized environment. Manoeuvring around in the dark takes a certain amount of skill, skill that most humans lack. But the cockroaches are fine. That's because cockroaches are clever, and they can learn many things, as seen through their ability to navigate mazes and memorize enough visuals to map their way back to shelter.

Turner, who we discussed in the last chapter for his pioneering work into the intelligence of bees, published his *Behaviour of the Common Roach (Periplaneta orientalis L) on an Open Maze* back in 1912. He observed female cockroaches that, as he wrote, 'grew accustomed to confinement and my presence' while he determined how long it took them to navigate a maze. And not just reach the end, but actually learn the route. Each time they failed, or when they had finished the maze, he would start them off again – and they became faster and faster at completing the task. He notes that initially the 'performance' takes between 15 and 60 minutes, but after a series of trials they could complete it within one to four minutes.

These clever cockroaches have relatively large brains in relation to other insects. Their brain is connected to a network of neurons that expands down the entire length of their body. And, in understanding this network, humans have been able to turn the field of neurophysiology on

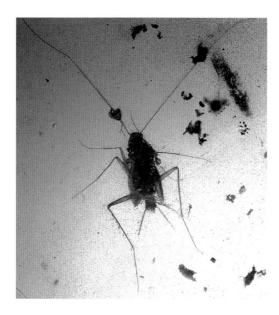

The world's oldest troglobite, the cockroach *Mulleriblattina bowangi* from the Cretaceous period, preserved in amber from the Hukawng Valley, Myanmar.

its head, as we learn just how quickly these animals are able to learn and adapt, and so in the process understand a little bit more about our own species. It's their extraordinary ability to learn that marks them out as major players in the exploration of the cellular mechanisms underlying animal performance, personality and behaviour.

Professor Stephen Simpson at the University of Sydney, Australia, has been studying the tiny nervous systems of insects. He grew up in in Australia and completed his undergraduate studies just up the road from work (albeit a very long one) at the University of Queensland, then completed his PhD at the University of London, UK, before moving to the University of Oxford. Through more than two decades of research whilst at the University, he has witnessed first-hand the extraordinary flexibility that enables them to adopt new behaviour in a matter of hours, or even minutes. As Simpson says, 'they're able to learn, they are able to be plastic to their behaviour in their environment – they are not just little robots that are programmed to do fixed things in fixed circumstances'.

So how do they learn exactly? Well, it's enabled by neuromodulation, by chemicals that run from one neuron to the next, affecting how many and to what extent each neuron operates, and thereby affecting how the entire system operates, how it perceives the world. It's like eating a fruit salad – by altering both the variety and the number of fruits in each mouthful, you can control the overall taste you experience in your mouth, and it will be different each time. By using just a small number of neurons flexibly, you can achieve great behavioural flexibility, and that's of huge adaptive significance for animals navigating their way in this world.

The idea that chemicals could play a role in the nervous system to affect the rest of the body is not a new concept and can be traced back to the eighteenth-century French physician Theophile de Bordeu (1722–1776). Bordeu was very much taken by the then accepted theory of vitalism – the presence of a vital spark or vital energy that separated living from non-living entities. Bordeu believed in particular that the glands, some of which we now know produce hormones, were dependent on this 'mystical vital force'. This speculative theory was in opposition to the prevailing ideas of iatrophysics or iatromechanics, simple mechanical explanations of the body's functions. Vitalism was eventually dismissed as pseudoscience, but Bordeu had begun to vaguely speculate about chemicals although he offered no data to support his musings.

The concept of chemicals affecting how the body worked was ridiculed throughout the nineteenth century, but it returned in a different form in the early 1930s thanks in part to a German biologist named Berta Scharrer (1906–1995). Born Berta Vogel to parents Johanna Weiss Vogel and Karl Philip Vogel, the latter was a successful judge and vice-president of the Federal Court of Bavaria. Along with her three siblings, Scharrer's childhood benefited from the cultural heritage of Munich, with much music and art and also an excellent education system. She attended a Gymnasium, which was the most advanced of Germany's three-tier secondary school system, and it was here that her interest in biology started, an interest she was to keep throughout her life. She set her sights on becoming a research scientist and in order to achieve her goal attended the University of Munich. That had been an

BORDEU.

The French physician Theophile de Bordeu, who speculated about the presence of 'vital energy' in living organisms.

astute choice. Going against the usual societal fashion by being a woman who attended the university, both her intelligence and aptitude led her to being supervised for her doctorate by physiologist Professor Karl von Frisch, who was researching bees. It must have been there, as Simpson states, that she realized, 'Little, tiny things were really remarkably clever'. But how these small-bodied, small-brained creatures achieved such proficiency – a bee's brain is less than 0.0002% of a human brain – must have been a burning question for her. She spent years researching the neuroanatomy of both bees and the tiny *Drosophila* for her doctorate, which she completed in 1930. But it wasn't all hard work – she met fellow student and vertebrate researcher Ernst Scharrer (1905–1965).

Ernst in 1928 had discovered what he called gland-nerve cells in the hypothalamus of fish – the brain's stabilizing and regulatory unit. Love blossomed (well that's hormones for you) which not only led to marriage but also an incredibly important working relationship. Scharrer was to comment that, 'An academic career at that time did not look promising at all for a woman. I must say that I could not have done what I have, if I had not been married to a biologist who gave me a chance to do my work.'

Historian Professor Matthew Cobb believes it was here that the seeds for her future career in the new field of neurosecretion – the storage, synthesis and release of hormones from neurons – were sown. For it was Ernst's 1928 discovery that neurons don't only send information by 'some kind of weird chemical electricity' but that they were also producing hormones. The only evidence he had at the time was the structures he could see in the nerves or as he wrote the 'colloid-like inclusions in the magnocellular preoptic nucleus of a teleost, the European minnow'. But his hypothesis was further supported by the rich vascularization that was apparent around the hypothalamus.

Scharrer and Ernst began working together, but as she herself witnessed, 'The idea that neurons may be capable of dispatching neurohormonal, or blood-borne signals, over distances, an activity previously associated only with endocrine cells, met with powerful resistance'. This was, after all, a controversial idea. Up to that point, the field of neurophysiology in the 1930s was still ignorant as to how nerve cells communicated. It wasn't even clear how a signal was transmitted down a nerve cell or how nerve cells sent their electrical signals from one cell to the next, let alone any ability they had to despatch so-called blood-borne signals.

A big argument had been taking place in science known as 'the war of the soups and the sparks', in which some researchers were adamant there was some kind of chemical message that passed from one neuron to another, whereas others, equally vociferous, thought it was a straightforward electrical signal. But Ernst's hypothesis was to strengthen the growing feeling it was some kind of chemical message that went from one nerve cell to another. He was not particularly interested in the actual movement of information down a neuron, but was captivated in determining whether neurons could produce hormones.

Scharrer began to look for evidence in various other species – a worm and a snail and saw to her surprise their neurons also possessed the structures Ernst had discovered. So immediately, she was able to provide supporting evidence for the controversial, and potentially ground-breaking, hypothesis her husband had developed. It raised the prospect that throughout much of the animal kingdom neurons were not only acting as wires, but were secreting hormones down those wires that were altering the behaviour and the development of organisms.

The Scharrers decided early on that they could best prove the widespread existence of neurosecretion by properly divvying up the animal kingdom. Ernst would continue to study vertebrates while Scharrer would take on the invertebrate world of insects. They were to prove a fantastic team, and it's this combination that would produce both the anatomical evidence and then actual experimental evidence leading to a holistic understanding of what was happening.

Scharrer began observing these neuron structures in her beloved bees and vinegar flies (*Drosophila*) and commented, 'One couldn't have foreseen the spectacular developments… It has been shown that the early observations were not artefacts or a figment of the imagination'. And although she didn't totally understand the process, Scharrer's first publications on the topic were in 1935 and 1936

Two of Scharrer's original study organisms – the depilatory sea hare (top) and the king ragworm (above).

studying different species of invertebrates – the molluscs called sea hares *Aplysia* (Linnaeus, 1767), and the king ragworm *Nereis virens* (= *Alitta virens*) (Sars, 1835).

Sharrer and Ernst often worked isolated from others to develop their newly developing theories. What must be a relationship goal for many a scientist was that they were able to bounce ideas off each other and reassure each other when doubts crept in. They were now both based at the University of Frankfurt, Germany, where Ernst had been appointed Director of the Edinger Institute for Brain Research. The rules of nepotism forbade Scharrer from receiving a salary, but she was given some research space, and along with a small stipend from her uncle they were able to survive. Be it under the auspice of 'nepotism' or just blatant sexism, the lack of salaried posts was to be a common theme throughout much of her life. But this was not her only problem, for the dark shadow of Nazi Germany was looming. A year before their marriage in 1934, the German parliament, the Reichstag, voted in favour of the Enabling Act, allowing Adolf Hitler to act without interference, a law that saw him able to rule as a dictator. Just over a week after that vote the infamous Law for the Restoration of the German Civil Service was passed that mandated the dismissal of all those defined as Jews from the civil service including academic staff. The Scharrers were Jewish, and although not themselves dismissed, they found it increasingly difficult to work under the new regime. Scharrer wrote, 'What the two of us were particularly opposed to was this senseless and immoral philosophy of the Nazis, the idea of racial superiority, anti-Semitism, genocide… We decided that it was impossible for us to be part of this system any longer'.

In 1937 they left for the USA, thanks to a Rockefeller fellowship for Ernst, which unknown to the authorities was to be a permanent relocation (via pitstops in Africa, the Philippines and Japan to collect study animals). Their first year was in Chicago and was the start of the couple taking a series of academic posts across America. Friend and colleague Dr George B Stefano, now at Charles University, Czech Republic, began working with Scharrer back in the 1980s, and recalls that while Ernst was able to secure paid work, Scharrer was forced to pursue her complementary research with Ernst for free. Stefano remembers that 'Being a woman gave her a tough time' but she still wanted to carry out meaningful work and so 'needed a research organism that was economical, easy to take care of, didn't demand a lot of people and was at the same time extremely meaningful'. Enter, the cockroach and to start off it was with 'pest' species *Periplaneta americana*. Well, they were easy to maintain, of good size and the centrally heated basement of Chicago University was crawling with them. Her lab or 'cockroach room' was a hive of activity (or should I say an intrusion of cockroaches – the amusing collective noun) It was located next to the ladies' loo, and people started to notice a smell – from the lab, not the loo. If you haven't smelt a cockroach, let me tell you they have a distinctive odour that some describe as musty or sour, or oily or sweet. The smell of cockroach can permeate the air, creating a greasy sensation. But with limited funds, Scharrer stuck with her new subjects.

There were plenty of cockroaches in Chicago, but when the couple moved to New York in 1938 the supply wasn't as good, a situation many modern-day New Yorkers would find ironic. But by 1940 she discovered a new cockroach in town, fresh off the boats – hello to the woodroach, *Leucophaea maderae* (Fabricius, 1781). The woodroach was originally from Africa, and had found its way to America via a shipment of primates from South America. It was big, and bold, much larger than the American cockroach, and thus much easier to dissect in Scharrer's 'operating theatre' – an excellent subject indeed.

In 1951 Scharrer published *The Woodroach* in *Scientific American* where she pays homage to 'This tough little creature… [which] survives all manner of experimental treatments. It is ideal for surgery, for it needs no aesthetic (it keeps perfectly quiet if fastened between the sheets of soft tissue paper)'. She also notes are how alike they are to the 'higher animals'.

When they moved next, to Cleveland, Ohio, the cockroaches came with them. Scharrer was still working unpaid, as a researcher and instructor. She wasn't even allowed to attend the departmental seminars to start with – only when she agreed to make the tea did they abate. Seven years later the couple were off again, again accompanied by the roaches, to the University of Colorado Medical school, and here Scharrer began her now classic experiments removing and transplanting tissue. It was through these experiments that she was able to prove the storage and release of hormonal substances from nerve cells. Not only that, but she showed how many of the hormones act quickly on neurons even at some distance away. The dynamic duo presented their case cautiously, with vividly colourful slides and plates showing these so called neural granules – some actually in the process of being secreted from their nerve cells. Her work was so highly regarded that in 1950 she organized a symposium in Paris and so really had to be given an academic title – finally she was made an assistant professor. But guess what, she was still unpaid. Despite this, she continued in her research, and completed her studies on the physiology of neurosecretory system of insects. Her research ideas were spreading, and her cockroaches were very quickly adopted as the physiologists' model organism of choice.

It can't be overstated, in terms of the big picture of the history of science, just what an extraordinary episode this was, as neurophysiology began to really take off. Scharrer and Ernst were doing groundbreaking work on the neurosecretional function of neurons, revealing a process

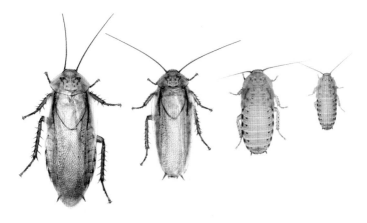

The woodroach – adult, pre-adult and nymphs.

that was hitherto unsuspected. At exactly the same time, other researchers were better understanding how neurons worked, in terms of transmitting signals around the nervous system.

By the 1950s this new knowledge had given rise to new drugs, both for medical purposes and for recreation, most famously lysergic acid diethylamide or as it was more commonly known LSD. It was a time in which the dual nature of neurons was beginning to be understood – electrical and chemical signals – and the work of Scharrer and Ernst tied in perfectly. The delay in being recognized was, according to Scharrer, a reasonable one, as she states, 'We had made bold claims and it is understandable that it took about twenty years of work for most people to accept these concepts'. But once this idea of neurosecretions was accepted, it would become the founding principle of a whole new discipline, the field of neuroendocrinology. Cobb highlights that it is because Scharrer showed neurosecretions in the invertebrates, and her husband showed it in the vertebrates, that we can say it appears across the whole of animal life. Their individual discoveries would not have been nearly as impressive as their combined ones.

Five years later they were off again, to the Albert Einstein College of Medicine in New York, and for the first time in her life, a salaried position for Scharrer. Ernst had been asked to set up the Department of Anatomy at this newly founded school and the dean offered Scharrer a full professorship as well. It being a new school he remarked, 'I know about the nepotism rule, but we are an entirely new school. We can do something a little progressive.' It was a progressive attitude that led to some very progressive work. In 1963, the pair published their classic work *Neuroendocrinology*, a staple for all students of the discipline. But with much sadness just two years later Ernst drowned during a break in Miami. A strong undercurrent took away Scharrer's husband and research partner, and her path was once more altered. With the same indomitable force that had powered her up to this point in her career, she continued to research right up until her death in 1995, still working at the age of 88.

While Scharrer's research helped revolutionize our understanding of nervous systems in insects, insects also paved the way to understand how nervous systems can drive rapid changes in behaviour. And one of the most powerful examples comes from biologist Professor Steve Simpson's favourite research tool, the desert locust *Shistocerca gregaria* (Forsskal, 1775). The desert locust shows extraordinarily malleable behaviour when changes occur to its environment. It can change reversibly from a shy and inconspicuous, solitary creature that flies at night to one that is highly conspicuous, flying by day and occasionally aggregating in vast numbers – with devastating economic effects. These two forms – the solitary and gregarious phases – are strikingly different in colour, physiology and behaviour. For instance the nymphs, those non-flying juveniles, of the solitary form are greener, to blend in with the vegetation for their docile lifestyle. The two forms can be bred in the laboratory and made to switch from one phase to the other and back, simply by raising them either in isolation or in a group. They have relatively few nerve cells in their brain, few enough to make it possible to observe the changes that take place during these phase transitions, to give insight into similar mechanisms that occur in more complex animals when they find themselves in new circumstances.

The key decision a locust must make is to join or avoid other locusts. Once this has been made, subsequent changes in physiology, body shape and colour follow, from the continuing presence or absence of other locusts. And these little winged insects can consume their own body weight in a day (2 g or 0.07 oz). This may not sound much, but multiply it by several million and you have a problem. From October 2003 to May 2005, many African countries were bombarded by the worst locust swarms for 15 years. Just one swarm in Morocco was 230 km by 150 m in size, containing an estimated 69 billion individuals. Twenty-three countries were impacted with the cost of fighting the swarms, estimated at $400 million. Changes in weather patterns result in many of these swarms and sporadic rain events are only predicted to increase in frequency as the impact of climate change becomes more evident.

The last major locust outbreak started in May 2018, after Cyclone Mekunu brought a biblical amount of rain to the Arabian Peninsula, followed in October by Cyclone Luban, which did the same. Locust numbers increased an estimated 8,000-fold as the remote location hampered access for the national locust teams. In 2019, the insects were on the move, up north to southern Iran and the Saudi Arabian interior and southwest to Yemen's interior. More rain, more movement, followed by more rain and yet more movement. Ethiopia, Somalia, Kenya, Uganda, South Sudan, Tanzania and for the first time since 1945 the Democratic Republic of the Congo, all hosted these unwanted arrivals.

All governments want to avoid the huge financial burden and massive loss of crops from these swarms, and so more research into what triggers them is imperative. While working as a post doc at the University of Oxford, UK, in the mid 1980s Simpson was funded by the UN to fly out to North Africa to study massive and destructive swarms of locusts, just as a decades-old insecticide dieldrin was banned. He took some back to his lab, where sensibly he kept them in small groups, while he began to investigate what it is about group living that triggers the switch. His early experiments determined that this Jekyll and Hyde transformation seemed to be caused not by sight or smell

but by physical contact between the locusts. To replicate the jostling action of a crowd, he used basic paintbrushes to tickle the legs of the locusts (with some sophisticated behavioural analysis thrown in) and was able to show that this physical action resulted in their physiological change. What's more, the change occurred in a matter of hours. Given the extraordinary speed of this change, Simpson and colleagues realized it could not be down to a rewiring of the nervous system – there wasn't enough time for that – but that the jostling had to be causing a chemical release that changed the strength of existing nerve connections. It was a chemical we know very well in our human world, serotonin, which influences our learning, mood, sleep and hunger. When these normally shy and solitary creatures undergo that big personality change upon encountering others, Simpson states that not only do they start seeking each other out and aggregating, but the adults develop enormous wings to enable them to amass into these migratory swarms. This is a type of

The solitary and gregarious nymphs of the desert locust.

polyphenism where multiple (in this case two) different phenotypes can arise due to changes in the environment. The nymphs are even different in colour with the solitary ones being greener to blend in with the vegetation.

So the changes in the nervous system that turned antisocial locusts into monstrous swarms was down to release of a chemical that also influences how we humans behave and interact – that great party hormone serotonin. Serotonin [5-hydroxytryptamine (5-HT)] is involved with many a function from modulating moods to vomiting – it gets you in the spirit and then deals with the impact of too much the next day! In a paper published in *Science* back in 2009, Simpson and colleagues reported there was a substantial but relatively short-lived (less than 24 hours) increase in serotonin when overcrowding occurred. They found that serotonin was both necessary for gregarization to happen and to sustain it. The fact the same chemical that causes a normally shy antisocial insect to gang up in huge groups is found in our brain is amazing. The effect of neurochemicals on nervous systems is what Simpson describes as 'utterly transformational in our understanding of biology'. He hopes that further research could reveal the specifics that are unique to the locust, that could one day prevent these swarms. Could drugs such as Prozac for us humans, which work by controlling serotonin, inspire something similar for our little but impactful munchers? Could the 'gang-up' effect of serotonin on locusts reveal something about our own human mob culture?

Scharrer brought the cockroach out of the basements and into the lab, and in doing so demonstrated the amazing adaptability these insects have, powered by a chemical process shared across the animal kingdom. It was a discovery that has led to even more discoveries about our world, and a reminder that we have more in common with cockroaches, and indeed every other animal on the planet, than we ever thought.

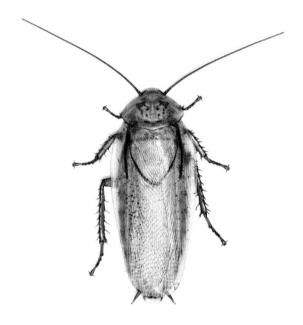

Concluding remarks

Every way you choose to think about the animals that dominate our planet they compel, astound and inspire. For millennia, we have been giving them names and studying their behaviour and from this has evolved a whole series of invaluable scientific endeavours that are proving as inspirational as when they were first described. Of insects, we have named but a few, let alone recognizing the many researchers who have obsessed over these fantastical species, some of whom were as ignored or ridiculed as their research subjects. Thankfully, perceptions have been overturned and more recently many scientists have been seeking inspiration from the natural world in the fields of medicine, space travel to fashion. Micro-technologists, systems engineers and biologists have much to learn from our smaller global brethren.

And so, what's next? Just think about how many species of insect surround us, and all the marvellous little tricks they have up their sleeves. Take a robberfly for example. There are over 8,100 species described to date from this one family of flies. And they are all quite extraordinary since these formidable predators have an arsenal of weapons and adaptations to help them hunt out their prey with amazing accuracy and efficiency. For these are venomous flies, one of the many neglected venomic species (what a term!) that we are just beginning to research. Unlike venom from the more well-known noxious insects the Hymenoptera, fly venom evolved independently multiple times and is throwing up unique venoms unknown to science. The family name has inspired the naming of some of these venom proteins with, to date, fifteen Asilidin protein families. The incredible fast acting nature of the venoms on their often highly aggressive prey (including on other highly dangerous species such as wasps, spiders and assassin beetles), may present new opportunities in anaesthetists' procedures. And to successfully hunt such active prey, they have evolved amazing eyesight – even the very small species have phenomenal vision.

The highly venomous *Asilus crabroniformis* Linnaeus, 1758 held by a slightly nervous Erica.

The genus *Holcocephala* Say, 1823 contains species smaller than a mosquito. I have come across these in Honduras and watched as they hunted out, often larger, prey. The eyes of flies (and all insects) comprise many facets that are basically individual photo units. HOw many there are depends on the species of insect and can range from zero to 36,000 (in the dragonflies). The robberflies don't have that many – since this species is approximately ten times smaller than your average dragonfly – so they have developed a type of zoom lens! At the centre of their eyes is a concentrated region of large lenses with some all-important associated light receptor cells (providing the all-important 'flash' for the images). These 6 mm flies can see, with great accuracy, over a distance of 0.5 m. They can accelerate until they are within 30 cm of their prey, when they 'lock on' to their targets and strike with incredible accuracy. That is me trying to spot and catch something that is initially a kilometre away. Can you imagine developing such small, lightweight but incredibly accurate range finders? We have so much to learn from these small marvels.

So this maybe the end of the book, but it is not the end – merely another instar - in our understanding of our fellow planetary inhabitants. Who knows what the foot of a beetle, the flight of a hoverfly or the faeces of a caterpillar will stimulate next? The history of insects transforming our world is littered with breakthroughs nobody saw coming, and as long as there are scientists with imagination and talent to think outside their disciplines, they will continue to inspire us.

Further reading

CHAPTER ONE: JUMPING JACK FLASH

Bergbreiter, S. et al. (2018), The principles of cascading power limits in small, fast biological and engineered systems. *Science*, 360 (6387).

Hopkins, G.H.E. and Rothschild, M. (1954), Rothchild Collection of Fleas. *Nature*, 173: 1204–6.

Kiick, K.L. et al. (2013), Resilin based hybrid hydrogels for cardiovascular tissue engineering. *Macromol. Chem. Phys.*, 214(2): 203–213.

Rothschild, M. (1964), Breeding of the rabbit flea (*Spilopsyllus cuniculi* (Dale)) controlled by the reproductive hormones of the host. *Nature*, 201: 103–104.

Sutton, G. and Burrows, M. (2011), Biomechanics of jumping in the flea. *J. Exp. Biol.*, 214(5): 836–847.

Tihelka, E. et al. (2020), Fleas are parasitic scorpionflies. *Palaeoentomology*, 3(6): 641–653.

CHAPTER TWO: MIGHTY MOUTHPARTS

Arditti, J., Elliot, J., Kitching, I.J. and Wasserthal, L.T. (2012), 'Good Heavens what insect can suck it' – Charles Darwin, *Angraecum sesquipedale* and *Xanthopan morganii praedicta*. *Bot. J. Linn.*, 169: 403–432.

Brożek, J. et al. (2015), The structure of extremely long mouthparts in the aphid genus *Stomaphis* Walker (Hemiptera: Sternorrhyncha: Aphididae). *Zoomorphology*, 134(3): 431–445.

Comparative Biomechanics and Evolution of Hawk Moth Proboscises https://cecas.clemson.edu/kornevlab/projects/

Endersby, J. (2010), *Imperial Nature: Joseph Hooker and the Practices of Victorian Science*. University of Chicago Press.

Endersby, J. (2016), *Orchid – A Cultural History*. University of Chicago Press.

Kornev, K., Monaenkova, D., Adler, P., Beard, C.E., and Lee, W.K. (2016), Butterfly proboscis as a fiber-based self-cleaning micro fluidic system. *Proc. SPIE*, 9797, p.979705.

Nishimotoab, S. and Bhushan, B. (2013), Bioinspired self-cleaning surfaces with superhydrophobicity, superoleophobicity, and superhydrophilicity. *RSC Adv.*, issue 3.

Pauw, A. et al. (2009), Flies and flowers in Darwin's race. *Evolution*, 63(1).

The Correspondence of Charles Darwin, Vol. 10. Darwin Correspondence Project. https://www.darwinproject.ac.uk/about/publications/correspondence-charles-darwin

Tonhasca, A. (2022), *Sticky Contrivances*. WordPress blog.

Wallace, A.R. (1867), *Creation by Law*. http://people.wku.edu/charles.smith/wallace/S140.htm.

Willis, K. and Fry, C. (2013), *Plants from Roots to Riches*. John Murray.

CHAPTER THREE: DROSOPHILA MELANOGASTRONAUTS

Anderson, D. and Brenner, S. (2008), Obituary Seymour Benzer. *Nature*, 451: 139.

Davenport, C.B. (1941), The early history of research with *Drosophila*. *Science*, Mar 28; 93(2413): 305–306.

Endersby, J. (2007), *A Guinea Pig's History of Biology*. Harvard University Press.

Greenspan, R.J. (1997), *Fly Pushing: The Theory and Practice of Drosophila Genetics*. Cold Spring Harbor Press.

Keller, A. (2007), *Drosophila melanogaster's* history as a human commensal. *Curr. Biol.*, 17(3), R77–R81.

Mohr, S. (2018), *First in Fly: Drosophila Research and Biological Discovery*. Harvard University Press.

Moore, M. et al. (1998), Ethanol intoxication in *Drosophila*: genetic and pharmacological evidence for regulation by the cAMP signaling pathway. *Cell*, 93 (6).

Stensmyr, M. et al. (2018), African *Drosophila melanogaster* are seasonal specialists on *Marula* fruit. *Curr. Biol.*, Dec 17; 28(24): 3960–3968.

CHAPTER FOUR: CYCLES OF CHANGE

Cobb, M. (2002), Malpighi, Swammerdam and the colourful silkworm: replication and visual representation in early modern science. *Ann. Sci.*, 59: 111–147.

Cole, F.J. (1951), History of micro-dissection. *Proc. R. Soc. London, Ser. B*, 138: 159–187.

Hammad, M. (2018), Bees and beekeeping in Ancient Egypt. *JAAUTH*, 15(1): 1–16.

Hassall, C. (2015), *Odonata* as candidate macroecological barometers for global climate change. *Freshw. Sci.*, 34(3).

Jorink, M.E. (2022), *Sibylla Merian and Johannes Swammerdam Conceptual Frameworks, Observational Strategies, and Visual Techniques*. Royal Netherlands Academy of Arts and Sciences.

Redi, F. (1909), *Experiments on the Generation of Insects*. https://ia601602.us.archive.org/0/items/experimentsongen00redi/experimentsongen00redi.pdf.

Swammerdam, J. (1669), *Historia Insectorum Generalis, ofte Algemeene verhandeling van de Bloedeloose Dierkens*. Meinardus van Drevnen, Amsterdam.

CHAPTER FIVE: BLOWFLY DETECTIVES

Anderson, G.S. (2020), *Biological Influences on Criminal Behavior*. 2nd edition. Taylor Francis, CRC Press and Simon Fraser University Publications.

Erzinclioglu, Y.Z. (1983), The application of entomology to forensic medicine. *Med. Sci. Law*, 23(1).

Hall, M.J.R. and Martín-Vega, D. (2019), Visualization of insect metamorphosis. *Phil. Trans. R. Soc. B*, 374: 20190071.

Malainey, S.L. and Anderson, G.S. (2020), Impact of confinement in vehicle trunks on decomposition and entomological colonization of carcasses. *PLoS ONE*, 15(4): e0231207.

Martín-Vega, D. (2017), Age estimation during the blow fly intra-puparial period: a qualitative and quantitative approach using micro-computed tomography. *Int. J. Legal Med.*, 131: 1429–1448.

CHAPTER SIX: DAZZLING DISGUISE

Behrens, R.R. (2009), Revisiting Abbott Thayer: non-scientific reflections about camouflage in art, war and zoology. *Phil. Trans. R. Soc. B*, 364: 497–501.

Dugdale, J.S. (1974), Female genital configuration in the classification of Lepidoptera. *N. Z. J. Zool.*, 1:2, 127–146.

Espeland, M. et al. (2015), A comprehensive and dated phylogenomic analysis of butterflies. *Curr. Biol.*, 28 (5): 770–778.e5.

Kjernsmo, K.M. et al. (2020), Iridescence as camouflage. *Curr. Biol.*, 30: 551–555.

Shell, H.R. (2009), The crucial moment of deception: Abbott Handerson Thayer's law of protective coloration. *Cabinet Magazine*, issue 33.

Waring, S. (2015), Margaret Foutaine: A lepidopterist remembered. *Notes Rec. R. Soc.* 69: 53–68.

CHAPTER SEVEN: THE ULTIMATE UPCYCLERS

De Sila, S.S. (2008), Towards understanding the impacts of the pet food industry on world fish and seafood supplies. *J. Agric. Environ. Ethics*, 21(5): 459–467.

McFadden, M.W. (1967), Soldier fly larvae in the United States North of Mexico. *Proc. U. S. Natl. Mus., Smithsonian Inst.*, 121(3569).

Smith, E.H. and Smith, J.R. (1996), Charles Valentine Riley the making of the man and his achievements. *Am. Entomol.*, 42(4).

Tomberlein, J.K. and van Huis, A. (2020), Black soldier fly from pest to 'crown jewel' of the insects as feed industry: an historical perspective. *J. Insects*, 6(1): 1–4.

Tomberlein, J.K. et al. (2009), Development of the black soldier fly (Diptera: Stratiomyidae) in relation to temperature. *Environ. Entomol.*, 38(3): 930–934.

Van Huis, A. et al. (2013), *Edible insects: Future prospects for food and feed security: FAO Forestry*. Paper 171. Food and Agriculture Organization, United Nations.

White, K.P. (2023), Food neophobia and disgust, but not hunger, predict willingness to eat insect protein. *Pers. Individ. Differ.*, Feb., vol. 202.

CHAPTER EIGHT: NAMIB FOG HARVESTER

Fernandez, J.C. et al. (2022), Optimizing fog harvesting by biomimicry. *Phys. Rev. Fluids*, 7, 033604.

Griswold, E. (1988), Obituary: Reginald Frederick Lawrence, 1897–1987. *J. Arachnol.*, 16(2).

Jiang, Y. et al. (2022), Coalescence-induced propulsion of droplets on a superhydrophilic wire. *Appl. Phys. Lett.*, 121(23).

Mitchell, D. et al. (2020), Fog and fauna of the Namib Desert: past and future. *Ecosphere*, 11(1).

Seely, M.K. (1979), Irregular fog as a water source for desert dune beetles. *Oecologia*, 42: 213–227.

CHAPTER NINE: BEE BRAIN INTELLIGENCE

Bridges, A., Chittka, L. et al. (2023), Bumblebees acquire alternative puzzlebox solutions via social learning. *PLoS Biol.*, 21(3): e3002019.

Chittka, L. and Rossi, N. (2022), Social cognition in insects. *Trends Cogn.*, 26(7).

Chittka, L. (2022), *The Mind of a Bee*. Princeton University Press.

Giurfa, M. et al. (2021), Charles Henry Turner and the cognitive behaviour of bees. *Apidologie*, 52: 684–695.

Jafferis, N., Helbling, E.F. et al. (2019), Untethered flight of an insect-sized flapping-wing microscale aerial vehicle. *Nature*, 570(7762).

Turner, C.H. (1911), Experiments on pattern-vision of the honey-bee. *Biol. Bull.*, 21(5): 249–264.

CHAPTER TEN: THE NERVE OF A COCKROACH

Pupura, D.P. (1998), *Berta V. Scharrer*. National Academies Press, vol. 74: 289–308.

Scharrer, B. (1951), The Woodroach. *SciAm.*, 186(6): 58–63.

Scharrer, B. (1992). *The Concept of Neurosecretion and Its Place in Neurobiology*. In: Worden, F.G., Swazey, J.P. and Adelman, G. (eds), *The Neurosciences: Paths of Discovery*. I. Birkhäuser Boston.

Sendi, H. (2020), Nocticolid cockroaches are the only known dinosaur age cave survivors. *Gondwana Res.*, 82: 288–298.

Simpson, S.J. et al. (2009), Serotonin mediates behavioral gregarization underlying swarm formation in desert locusts. *Science*, 323(5914): 627–630.

Smith, D.B. et al. (2016), Exploring miniature insect brains using micro-CT scanning techniques. *Sci. Rep.*, 6, 21768.

Turner, C.H. (1912), Behaviour of the Common Roach (*Periplaneta orientalis* L.) on an open maze. *J. Univ. Chicago*, 348–365. https://www.journals.uchicago.edu/doi/pdf/10.1086/BBLv25n6p348.

Index

Picture credits

Acknowledgements

Writing this book would not have been possible without the help of the many entomologists, historians, librarians, engineers, and medics who gave up their time and provided valuable source material for our BBC Radio 4 series *Metamorphosis*, from which the stories in this book evolved. In particular, we'd like to thank to Peter Adler, Gail Anderson, Max Barclay, Sharmila Battacharya, Sarah Bergbreiter, Lars Chittka, Matthew Cobb, Jim Endersby, Martin Giurfa, Martin Hall, Andrea Hart, Chris Hassell, Farrell Helbling, Kristii Kiick, Jeff Kirkwood, Ian Kitching, Karin Kjernsmo, Richard Lane, Duncan Mitchell, Stephanie Mohr, Andrew Parker, Mary Seely, Steve Simpson, George Stefano, Greg Sutton, Jeff Tomberlin, Grace Touzel, Katharina Unger, Daniel Martin Vega, David Waterhouse, Don Webber and Keiran Whittaker. And also thank you to the millions of insects themselves – truly inspirational.

Thanks to the Museum's Publishing team and the Museum's Diptera Team, our partners for their advice and peerless grammar pedantry (and gin), and Alfie and Rags and Ruby for demanding awe-inspiring cuddles/walks between bouts of writing.